SpringerBriefs in Electrical and Computer Engineering

For further volumes:
http://www.springer.com/series/10059

Sleiman Bou-Sleiman · Mohammed Ismail

Built-in-Self-Test and Digital Self-Calibration for RF SoCs

Springer

Sleiman Bou-Sleiman
Analog VLSI Lab
The Ohio State University
Columbus, OH, USA
bousles@ece.osu.edu

Mohammed Ismail
Analog VLSI Lab
The Ohio State University
Columbus, OH, USA
ismail@ece.osu.edu

ISSN 2191-8112   e-ISSN 2191-8120
ISBN 978-1-4419-9547-6   e-ISBN 978-1-4419-9548-3
DOI 10.1007/978-1-4419-9548-3
Springer New York Dordrecht Heidelberg London

Library of Congress Control Number: 2011937576

Printed on acid-free paper

Springer is part of Springer Science+Business Media (www.springer.com)

To my loving family

        Sleiman Bou-Sleiman

To the memory of my father,
Ismail El-Naggar

        Mohammed Ismail

Preface

Single-chip radio systems, or Radio Frequency System-on-Chips (RF SoC), have become increasingly popular in recent years driven by the many aspects and intersections of technology, market demands, and consumer needs. Consumers now expect flawless and seamless communication capabilities in their increasingly connected world. Satisfying these demands is contingent on the abilities of scientists and engineers to continually advance the state-of-the-art in microelectronics processes and circuit designs. On the technology side, the improvements in silicon MOS have enabled high performance and highly integrated circuits, taking CMOS from a purely digital to a mixed-mode technology. For companies, the ability to provide platforms and systems built around a common technology, in smaller form factors and with more horsepower is a critical aspect of their survival in a deeply competitive market. The successful merge of the latest technology with the best design practices is the catalyst to first time design success.

The continual physical shrinking of device dimensions is allowing for more integration between the previously segmented digital logic, memory, analog, and radio frequency domains. While this co-existence may indicate a cost reduction on paper, in reality it might well turn out in the red. The smaller device sizes, while faster, are becoming increasingly unreliable. Although able to meet the performance requirements for high-speed analog and RF, the devices are not guaranteed to always run at their typical sweet spot. The drifts from the optimal operation are due to many factors related to the silicon process and its response to changes in voltage and temperature, or what is collectively named *PVT* (Process, Voltage, Temperature) variations. These variations are a problem in all the integrated domains of the chip; however, RF and millimeter-wave (mm-wave) circuits fail, in a more disproportionate manner, at sustaining proper operation over PVT. The reason being that RF devices, unlike digital circuits, do not function exclusively as on-off switches at either edges of operation but also exercise all the continuous states in between, and at very high frequencies. This makes them more prone to performance degradations and loss of yield when fabricated, in contrast to digital chips that can achieve near perfect yield. Putting both RF and digital together on a single chip, the hybrid system obviously inherits the lower yield, negating all the integration advantages.

Therefore, the RF portions, in a sense, represent the SoC's Achilles' heel; in essence, an overly powerful and densely integrated chip can be made useless by a smaller underperforming portion of the chip.

The ultimate goal is to increase the functional yield of the RF blocks by actively maintaining them in their optimal operating region. This proves to be a non-trivial task, as the operating conditions of the system at all times need to be known. While fabrication testing is one way to test how a chip performs after production, it cannot be all-inclusive of all operating conditions. Moreover, it is quite costly to fully verify each single chip rendering the validation task quite prohibitive. A solution would be to build self-testing, and eventually self-healing, systems. However, this demands a shift in design paradigms to include testing, early on, in the design phase, or what is dubbed as Design-for-Testability (DfT). While DfT's primary goal is to ease external testing of complex chips, an additional upgrade is highly desirable: the integration of the full testing functionality on-chip.

Built-in-Self-Test (BiST) techniques have already established themselves in the validation of digital blocks but are now becoming an increasingly active domain of research and development in RF. The notion of migrating RF test functionality to inside the chip brings us one step closer to cognitive-like radios. Self-awareness in RF systems is therefore a product of efficient on-chip test generation and result acquisition and interpretation. If RF/mm-wave blocks and systems can test for, and extract, their performance, then the ability to calibrate and cancel discrepancies can also be built into the system. Hence, Built-in-Self-Calibration (BiSC) can be layered on top of BiST to result in auto-correcting RF impairments at the block and system levels.

In this book, we discuss both BiST and BiSC in the context of RF SoCs. In Chapter 1, we describe CMOS' roadmap towards RF and mm-wave capabilities. The beneficial and adverse effects of technology scaling are highlighted showing the possibilities of increased integration as well as the problems associated with decreased device and circuit robustness. Chapter 2 describes the basics of communication systems and transceivers, while highlighting the most critical performance issues. System- and block-level metrics are presented along with RF built-in-testing schemes to reduce the costs of production testing. In Chapter 3, we express the requirements for building efficient true self-test mechanisms using on-chip resources not only as value-added elements but also as necessary components for successful first-pass success of RF and mm-wave SoCs. Simple additional circuits for test are therefore desirable; however, on-chip testing of radio frequency signal needs special attention, as these signals' properties are not easily interpretable. Hence, an efficient RF detector is presented with different implementations covering the RF and mm-wave spectra. In Chapter 4, the detector is used under different built-in-test schemes for parametric extraction of RF blocks. The self-testing of metrics such as gain, linearity, and quadrature mismatch is described with example test-benches and circuits. Chapter 5 takes the proposed on-chip test implementations and uses them to aid in the development of calibration techniques. These techniques aim at leveraging the strengths of the more robust parts of the system to cover up the weaknesses of the others. Therefore, the digital domain is fully exploited to augment the capabilities of RF circuits by providing them with the digital notion of programmability: an added degree of flexibility and tunability with the goal of enabling performance steering capabilities. Examples of RF Built-in-Self-Calibration using DSP-driven approaches are briefly highlighted and shown to re-adjust a failing circuit's operating conditions.

This book is intended for RF design engineers, system-on-chip design engineers as well as graduate students and researchers in the field. The material strives to present an approach and a description of a process that fits perfectly into the premise of, and promise of, highly performing first-time-right design of RF SoC moving into the nanometer regimes.

The work in this book has its roots in the PhD work of the first author at the Analog VLSI Lab at the Ohio State University. Both authors would like to acknowledge the support of colleagues at the Analog VLSI Lab and Electroscience Laboratory at the Ohio State University.

Columbus, OH   Sleiman Bou-Sleiman
                                                                                 Mohammed Ismail

Contents

List of Figures

List of Tables

Chapter 1
Introduction and Motivation

After having established itself as the dominant process for digital microelectronics, the past decade saw the emergence of CMOS as a viable process for Radio Frequency (RF) blocks. The advances in semiconductor technologies allowed for new long-sought-after opportunities for high scale integration of complete systems on a single chip, or System-on-Chip (SoC). The ability to successfully embed RF, analog, and digital portions of radio systems into a fully integrated solution brought about many new and interesting products and applications (Fig. 1.1). As much as this has opened up possibilities, such high integration is not without its shortcomings.

With every new boundary overcome in producing the next generation of advanced process technologies, we begin to observe a new set of challenges and impediments. This is more visible with analog and RF circuits. Inasmuch as digital circuitry and memories benefit from miniaturization, analog and RF circuits' behavior starts to show increased susceptibility to deviate from optimal performance, therefore negating any gains made possible by their smaller sizes. This causes analog and RF circuits to suffer from low yield due to process, supply, and temperature (PVT) variations and therefore require several expensive silicon design and manufacturing cycles to meet their specifications. Also, when co-existing with a mass of fast-switching digital logic on the same substrate, they are inherently subjected to additional noise coupling. These issues have put increased limitations and restrictions on the design of RFIC blocks in platform baseband SoCs. To tackle the problems associated with process shifts, parasitic elements, and changing operating conditions, recent efforts concentrating on novel design techniques have emerged with the goal of minimizing the yield loss in radio SoCs. To the industry, this essentially translates to reduced engineering costs, faster product development, and faster time to market.

To reduce yield loss due to variability, post-silicon calibration is necessary to compensate the performance degradation. Calibration of RF blocks, however, is not an easy and straightforward task – it requires enhanced test mechanisms to be incorporated into the design of the system. Responding to these difficulties, Design for Testability (DfT) within RF systems has caught ground and enabled the implementation of Built-in-Self-Test, or BiST. This emerging field of research and development

S. Bou-Sleiman and M. Ismail, *Built-in-Self-Test and Digital Self-Calibration for RF SoCs*, SpringerBriefs in Electrical and Computer Engineering, DOI 10.1007/978-1-4419-9548-3_1, © Springer Science+Business Media, LLC 2012

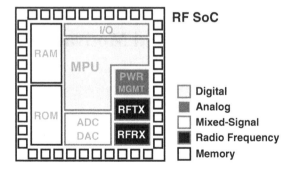

Fig. 1.1 System-on-Chip: Single-chip radio systems with RF, analog, memory and digital

opened the way for circuit self-awareness. Self-aware circuits will ideally not only be able to test themselves – detecting their operating conditions – but also correct for deviations through self-calibration procedures, to revert to optimal performance under any operating point. This vision of self-healing RF circuits can then only be achieved with new methodologies for pre-silicon designs incorporating robustness enhancement at their center.

In this chapter, we briefly describe the factors enabling CMOS RFICs and radio SoCs. We discuss the effects of technology scaling, variability in the nanometer regime, and some of the costs associated with integrated circuits. This will provide scope to the rest of the book which aims at presenting a description of the challenges in design and test (Chap. 2) as well as suggesting solutions towards building robust first-pass RF SoCs through efficient self-test (Chaps. 3 and 4) and self-calibration (Chap. 5).

1.1   The Need for Robust RF and mm-Wave ICs

As nanoscale CMOS pushed its operating boundaries beyond the digital into the high-frequency analog and RF and more recently the mm-wave realm, the focus has been on how to increase the yield of these ICs, standalone and in heterogeneous systems. The ultimate goal is then to achieve first-pass success through first-time-right design techniques and methods for radio SoCs. To understand the need for building robust ICs, we briefly discuss the driving forces and enabling factors for CMOS radios in terms of integration trends and device scaling. Like most technologies, highly-integrated CMOS SoCs will only make sense if they provide an economically viable solution. For that, the cost implications of the entire cycle spanning design, manufacturing, and test plays into the overall equation.

1.1.1   Integration Trends in CMOS

In 1965, Intel cofounder Dr. Gordon E. Moore penned a formulation describing his observations regarding the near-exponential increase in semiconductor component integration and extrapolation of the trend into the then forthcoming decade. His statement, Moore's law, as it came to be called, is in essence a remark on the cost of

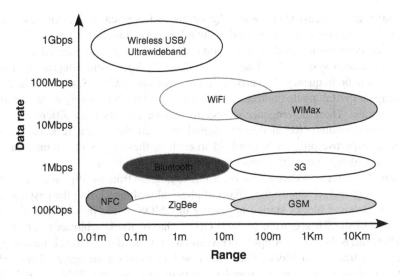

Fig. 1.2 Data rates and ranges of various wireless standards

integration: "The complexity for minimum component costs has increased at a rate of roughly a factor of 2 per year" [1]. Revised later to the more common notion of doubling transistor densities every 2 years, the trend held up for over half a century with effects not only in the density of transistors but also in related computational technologies, such as memory capacities, processing power, image sensors, and others.

Semiconductor companies have adopted Moore's law as a benchmark and milestone in their technology advancements and goals. This has spawned novel technologies and sparked innovations in trying to keep with the ever increasing exponential goals. When photolithography techniques could not surpass the limitations of light wavelengths to create smaller and smaller transistors, breakthroughs in extreme ultraviolet (EUV), double-patterning, and maskless techniques have ensured the survival of the trend to minimum sizes down to 22 nm and 14 nm at the time of writing [2, 3]. Back in 1971, the first general purpose microprocessor, the Intel 4004, held approximately 2,300 transistors at 10 µm length with processing power around 0.07 Million Instructions per Second (MIPS). In 2011, Intel's own Core i7 processor counts 915 million transistors at 32 nm length capable of close to 160,000 MIPS. Today's systems continue the extreme scaling trend and are expected to increase their densities and computational capacities in the near future, albeit with more innovations as we close in on physical atomic limitations.

As people's mobility increased so did their demand that computational powers travel with them. The surge in integrated electronics has catalyzed yet another revolution: the wireless revolution. From pagers in the mid 1970s to the 1980s and 1990s' mobile phones then notebook computers and more recently smart-phones and also stretching to myriad other applications such as body-implantable wireless medical devices or even smart utility metering, the wireless revolution enabled prompt and seamless communication and made ubiquitous connectivity a staple of our daily lives. Figure 1.2 shows a subset of the prevalent wireless standards, their data rates and ranges, showcasing the breadth of applications from near field connectivity to long distance 3G and 4G wireless communications.

Digital computation and wireless communication have then merged to create single integrated systems for data transmission. The successful integration of the digital and radio frequency (RF) capabilities in plain silicon CMOS, a process heavily focusing on digital, made building platform System-on-Chips a very attractive and viable option. These heterogeneous systems include circuits from different signal processing domains – RF, analog and digital. As such, the design, test, and manufacturing expertise previously confined to each of these types of circuits, has to merge accordingly presenting new challenges and possibilities.

CMOS' circuit boundaries have been able to migrate from the digital to the high-speed analog and RF due to the effects of physical scaling. During the past decade or so, transceiver blocks started appearing in CMOS as the process was able of operation in the RF spectrum (0.4–10 GHz). One of the last blocks to enter the CMOS realm is the Power Amplifier (PA) and with it added to the mix, little remains in terms of transceiver blocks that cannot be integrated on the same silicon substrate. Single-chip solutions for popular standards such as WiFi, WiMAX, and LTE are now available and are designed to meet the growing demands for high speed data, enhanced applications, and high levels of mobility.

More recently, as higher speeds are achievable in CMOS, the mm-wave spectrum has gained interest for commercial exploration, especially for short-range high-throughput applications. While CMOS has never been the process of choice for extreme high frequencies, its promise for increased integration and reduced cost make it the strongest contender to cover non-niche applications steered towards the mass consumer markets, e.g. WirelessHD (IEEE 802.15.3c) [4].

1.1.2 CMOS Scaling Effects

The effects of scaling were first observed by the early researchers in the semiconductor field where they noticed benefits in nearly all aspects of a shrunken transistor. Physical process scaling has steadily continued to bring in advantages in speed and chip area – two very sought after metrics for future investment in any technology. However, disadvantages also abound, more so with the newer scaling efforts as further size reductions are now less possible.

Table 1.1 shows the more recent and expected CMOS scaling trends [5]. As the transistor's length is decreased, we see an inversely proportional increase in the speed of the MOS device ($1/L_{gate}$). The reduction in area, on the other hand, is proportional to L_{gate}^2. Supply voltages are also reduced to lower the power consumption coupled with a decrease in the threshold voltages (V_{th}). This reduction in threshold voltage is necessary for proper operation under lower voltage supplies but has the disadvantage of exponentially increasing the sub-threshold leakage, which is starting to account for 50% of the total power consumed in modern embedded systems [6]. As further scaling of planar CMOS is unsustainable, new transistor

Table 1.1 CMOS technology roadmap

Year	2003	2005	2010	2013	2016
Gate length (nm)	65	45	32	22	15
f_T (GHz)	183	264	280	400	570
T_{ox} (nm)	1.6	1.5	1.5	1.2	1.1
V_{th} (V)	0.18	0.15	0.11	0.1	0.1
V_{dd} (V)	1.2	1.2	1.1	1.0	0.95

Fig. 1.3 Transistor cut-off frequencies for different processes and geometry nodes

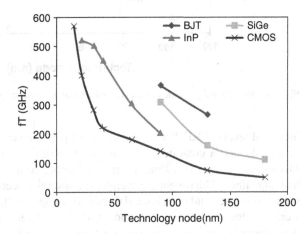

structures are under study for the 14 nm nodes and beyond, such as FinFETs [7] and three-dimensional tri-gate structures [8].

With each advanced process node, faster transistors became possible that not only increased the digital circuits' operating frequency but also made it possible to build RF and mm-wave circuits spanning the low to tens of GHz range. While the latter types of circuits were typically manufactured in SiGe and GaAs due to their superior RF performance (noise, speed, etc.), CMOS device f_T has steadily increased with each generation making it capable of handling high-frequency wireless. Figure 1.3 compares silicon CMOS with the other process technologies that occupy the RF and mm-wave space. CMOS has always led the pack in scaling with the other technologies trailing by approximately two generations [9]. For a process technology to be capable of a certain application, its f_T should exceed the operating frequency by tenfold, for small-signal operation, or twice, for oscillation and power transistors. As seen in Fig. 1.3, nanoscale CMOS is well within the RF spectrum and closing in on the mm-wave range with each process generation.

For CMOS to operate RF devices, it needs to exhibit certain figures of merit such as gain, noise figure, output power, and linearity among others. As many of the changes to bulk CMOS are done to implement faster digital circuitry, this does not necessarily translate to better analog and RF metrics. Variations are much more pronounced in deep-submicron and nanometer devices thereby manifesting in

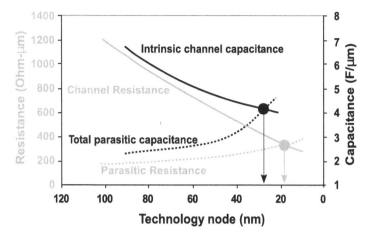

Fig. 1.4 Intrinsic and parasitic channel capacitance and resistance per technology node

increased levels of RF distortion and loss of performance [10, 11]. Analog and RF circuits depend a great deal on matching between devices, e.g. differential pairs. Mismatches due to variations in gate dielectric, random dopant fluctuation, and line-edge/line-width roughness, weigh in on several aspects of the device's behavior, such as sub-threshold currents and threshold voltages [12]. Undesired parasitics that were once too small to notice are now sufficiently large with respect to the intrinsic device parameters – and are expected to increase further, as shown in Fig. 1.4 for channel resistance and capacitance [13].

Another type of parasitics, interconnect parasitics, plays a pivotal role in the performance of transceivers and SoCs, especially when routing high frequency signals, such as RF and high-speed I/O. With highly integrated RF SoCs, the number of interconnect layers has increased to allow for sufficient connectivity for the embedded complex circuitry – typical submicron processes can have up to ten metal layers, with top metals suitable for RF passives and signal routing. So, apart from the device parasitics that limit the performance of individual circuits, SoC performance is also limited by the interconnect between, and beyond of, the individual blocks. These limitations arise from factors such as RC delay, IR drop, and cross-talk [14]. With parasitic effects amplified at very high frequencies such as in mm-wave applications, metal wires' passives like sheet resistance, coupled capacitance to nearby metals and substrate, and mutual inductance have to be embedded and modeled into the design process [15, 16]. Their effects on system parameters cannot be ignored, especially with heterogeneous systems having fast switching digital circuits and precision analog and RF on the same silicon substrate [17].

Process variations and parasitics skew the performance of analog circuits and force designers to include substantial margins to reduce and try to eliminate yield loss. Over-design usually comes at the expense of non-optimal power solutions, paramount to over-kill. Figure 1.5 shows the various forces affecting circuit design and reducing yield with their adverse variability. It is quite important to have accurate

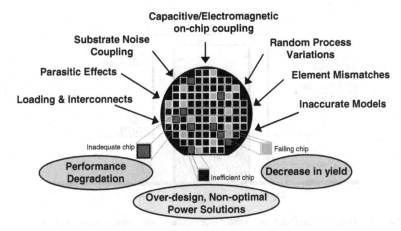

Fig. 1.5 The many contributors to variability in nanometer CMOS technologies

Table 1.2 Threshold voltage variability

Technology	180 nm	130 nm	90 nm	65 nm	45 nm
σV_{th}	5.8%	8.2%	9.3%	10.7%	16%

models that capture these effects at early stages of the design as well as precise control over passives. However, designers have to take special care and invest more design effort to ensure performance robustness from the RF and analog parts. Apart from the intrinsic changes in the process, environmental variations in the operating conditions, such as temperature and supply voltage changes, have an equally important effect on the circuit behavior. Collectively called PVT variations (for process, voltage, and temperature), they form a set of corner cases which designers have to account for. NMOS and PMOS devices can either be fast or slow, resistance and capacitance values can vary by several tens of percentage points, and each device has its own temperature gradient – any combination of these cases and others form a set of corners that can manifest in a silicon run, even within the same wafer. With newer processes, the number of possible corner cases has dramatically increased forcing designers to resort to substantial over-design. Therefore, solutions are needed to mitigate performance degradations through novel techniques in process, design, and layout. And in order to reduce the silicon spins and achieve first-time-right design, these best-practice techniques can be augmented with self-test and self-calibration schemes – both topics discussed in later chapters.

The advantages enabled by scaling can be easily shadowed by the disadvantages coming from the increased variability. Table 1.2 lists the increasing threshold voltage variations with decreasing geometries [18]. The superposition of several such variations leads to a very wide distribution of device performances, to the first degree, and eventually circuit/block performance uncertainty. Moreover, as we look at newer wireless standards, they impose more stringent system performance

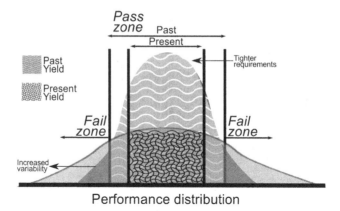

Fig. 1.6 The effects of variability and tighter specifications and requirements on yield

requirements effectively narrowing the pass/fail boundaries. These complimentary effects result in more yield losses (Fig. 1.6) and subsequently more design, test, and fabrication investments.

Digital designs have so far been able to keep a very high yield, owing to their discrete nature of operation, something RF and analog blocks cannot boast. The loss of performance from the latter blocks will eventually set the pass/fail limit of the entire chip – rendering completely flawless digital designs useless with a single faulty, or sub-performing, RF block in the SoC.

1.1.3 Cost Factors

As mentioned previously, Moore's law's original statement is an observation regarding cost. Figure 1.7, Moore's original redrawn sketch from [1], shows that the cost of integrating a single transistor was decreasing with time. And ever since, the number of transistors that can be integrated at minimum cost has been increasing exponentially making the cost of a single transistor extremely cheap, as seen in Fig. 1.8. However as the latter figure also shows, silicon foundry tool costs are on the increase translating to more expensive design fabrication for the IC firms. Driving the costs up is the foundries' continued investment in research and development to push the limits of the technology, helping them create larger wafers, increasing the yield, and integrating more devices onto silicon [19]. All these goals attempt to decrease the cost of integrating a transistor – however, the probability of that transistor being a faulty device is increasing with each denser process node. This necessitates proper and accurate device testing at the production level. Testing costs have generally remained constant but with the decrease in integration costs, testing cost per transistor has become comparable to the latter. To the IC design firms, fabrication costs are quite expensive with typical nanometer CMOS mask sets cost reaching a couple million US dollars [20].

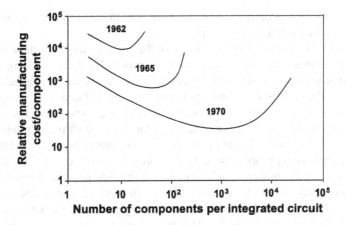

Fig. 1.7 Moore's observation on the relative manufacturing cost and number of integrated components [1]

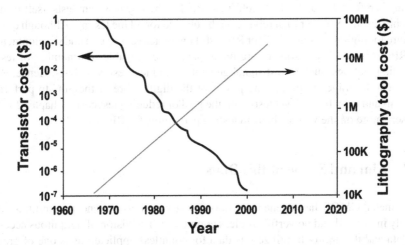

Fig. 1.8 The opposing trends in transistor cost and lithography tool cost [13]

Fabrication cost is but one of the many costs associated with the lifetime of an IC product. Prior to mass production, several iterations of design, manufacturing, and testing might be necessary to reach the desired performance of an SoC. All of these phases entail sizeable costs. A single design cycle can take upwards of 1 year, between setting system specifications, system design, then circuit design, layout, and post-layout verification – with multiple possible revisions to any of the previous. During that primary design cycle, costs include salaries, software and design tools. For any subsequent refinement cycle, non-recurrent engineering (NRE) costs start chipping from the possible revenue, in actual cost and time. Of extreme importance in a very competitive semiconductor market is the value of time. A balance needs to exist between reducing any delays in time-to-market to maximize profits over the product's life cycle, and rushing that product to the market without proper verification and testing. Defective products that reach the end customer can cost more than 10,000 times their selling price to replace [18].

Testing RF SoCs is not a trivial task given the complex nature of these systems. Based on the issues discussed so far, it becomes evident that RF SoC testing needs to achieve high accuracy while reducing costs. Testing costs are in themselves also high, both at the prototype and mass production stages. In the first, IC design firms use laboratory setups with a handful of expensive test equipment performing multiple types of tests to assess the performance of the designed chips. At production, a more automated testing solution in the form of Automatic Test Equipment (ATE) exists to quickly verify proper system functionality. In either cases, RF SoCs present a challenge in testing as they include a large number of test points and possible faults that are not easily accessible from the outside. To this end, circuit designers have recently resorted to embed test-enablers and even complete test macros on-chip in addition to the main system blocks. This is the basis for the Design-for-Testability (DfT) approach. With DfT, some of the testing can be migrated on-chip and test results can be interfaced and read from off-chip using low cost equipment. Embedded testing is referred to as Built-in-Self-Test (BiST) whereby a system tests itself using on-chip resources rather than relying on the outside for all the testing. Although quite mature for digital testing, BiST for RF is still under active research and development. In RF SoCs, on-chip resources are plenty and hence capable of a number of test-specific functions, such as test signal generation and test result analysis. Hence, RF BiST can leverage the processing power of the digital core of the SoC to perform system- and even block-level tests with the goal of reducing test costs. Chapter 2 will present more on the various built-in testing paradigms for RF SoCs.

1.2   Aim and Scope of this Book

Nanometer CMOS has opened the door for a new horizon and possibilities for highly integrated and powerful wireless computing. The vision of ubiquitous access to data and the ability to utilize this data for countless applications is one of great interest to consumers, companies, and academia.

But for these technologies to reach maturity they have to surpass many challenges in their design, manufacturing and test. Only when cost-effective can a certain application make its way to the mass market. Therefore, IC designers have to overcome many of the limitations that prevent the successive push towards more integration of computing powers and wireless connectivity. As the current state of process technologies allows for such capabilities to be implemented, designers are going to lengths to make coexistence of powerful data processing and wireless data communication stable, robust, and effective, in in-field operation and more importantly in pre-deployment cost. Chapter 2 describes the building blocks of radio transceivers, system and block metrics, and the more suitable architectures for SoCs. More importantly, we also discuss the shortcomings and limitations of the radio system's building blocks and describe built-in test techniques to quantify their performance.

The challenge then is how to benefit from the advances in various design and test domains to achieve first-time-right RF and mm-wave radio SoCs. As we move to atomic scales, robustness is severely degraded and has to be artificially augmented. This augmentation is only possible with novel designs built around test and calibration. Novel methodologies encompassing state-of-the-art techniques and circuitry are needed. The ultimate goal is to instill a heightened sense of cognition, along the lines of artificial intelligence, into radios and systems – steering their subpar performance up to specifications through the clever use of available resources. Flexible radios and smart systems will have to embed built-in-self-test and built-in-self-calibration (BiST and BiSC).

Design for testability and tunability is a hot field in research and development. The enabling circuits for testability are still being developed for RF and mm-wave; to this end, we describe one such effective and low-overhead circuit in Chap. 3. Many challenges exist in making on-chip detection reliably accurate, especially when dealing with very high frequency signals. Self-test examples are investigated in Chap. 4. Coupled with the clever use of robust resources to heal the less immune circuits through tunability, Chap. 5 discusses the use of analog and digital calibration to enable first-pass silicon success, cutting costs and eventually freeing funds for further technological development.

References

1. Gordon E. Moore, "Cramming More Components onto Integrated Circuits," *Electronics*, vol. 38, no. 8, pp. 114–117, April 1965
2. D. McGrath, M. LaPedus, "Analysis: Litho world needs a shrink," *EE Times*, 3/14/2011 [Online]. Available: http://www.eetimes.com/electronics-news/4213996/Analysis--Litho-world-needs-a-shrink-
3. M. LaPedus, "Globalfoundries, TSMC square off in litho," *EE Times*, 3/1/2011 [Online]. Available: http://www.eetimes.com/electronics-news/4213679/Globalfoundries--TSMC-square-off-in-litho/
4. WirelessHD, http://www.wirelesshd.org/
5. The International Technology Roadmap for Semiconductors: 2010 [Online]. Available: http://www.itrs.net/Links/2010ITRS/Home2010.htm
6. B. Al-Hashimi, Ed., *System-on-Chip: Next Generation Electronics*. IEE Press, May 2006
7. M. Guillorn, "FinFET Performance Advantage at 22 nm: An AC perspective," *VLSI Technology, 2008 Symposium on*, pp. 12–13, Aug 2008
8. R. S. Chau, "Integrated CMOS Tri-Gate Transistors: paving the Way to Future Technology Generations," *Technology@Intel Magazine*, Aug 2006
9. A. J. Joseph, D. L. Harame, B. Jagannathan, D. Coolbaugh, D. Ahlgren, J. Magerlein, L. Lanzerotti, N. Feilchenfeld, S. St. Onge, J. Dunn, E. Nowak, "Status and Direction of Communication Technologies - SiGe BiCMOS and RFCMOS", *Proc. IEEE*, vol. 93, pp. 1539–1558, 2005
10. R. van Langevelde, et al., "RF-distortion in deep-submicron CMOS technologies," *Proc. Int. Electron Devices Meeting*, pp. 807–810, 2000
11. C. Choi, et al., "Impact of poly-gate depletion on MOS RF linearity," *IEEE Electron Device Lett.*, vol. 24, no. 5, pp. 330–332, 2003

12. K. Kuhn et al., "Managing Process Variation in Intel's 45 nm technology," *Intel Tech. J.*, vol. 12, pp. 93–109, 2008

13. S. E. Thompson, S. Parthasarathy, "Moore's law: the future of Si microelectronics," *Materials Today*, Volume 9, Issue 6, pp. 20–25, June 2006

14. T. Kunikiyo, et al., "Test structure measuring inter- and intralayer coupling capacitance of interconnection with subfemtofarad resolution," *IEEE Trans. Electron Devices*, vol. 51, no. 5, pp. 726–735, 2004

15. M. El-Desouki, et al., "The impact of On-chip Interconnections on CMOS RF Integrated circuits", *IEEE Electron Devices*, vol.56, no.9, Sept 2009

16. B. Kleveland, et al., "High-frequency characterization of on-chip digital interconnects," IEEE J. Solid-State Circuits, vol. 37, pp. 716–725, 2002

17. A. Helmy and M. Ismail, *Substrate Noise Coupling in RFICs*. Springer, 2008

18. C. Chiang and J. Kawa, *Design for Manufacturability and Yield for Nano-scale CMOS*. Springer, 2007

19. B.P. Wong, et al., *Nano-CMOS Design for Manufacturability: Robust Circuit and Physical Design for Sub-65 nm Technology Nodes.*" Wiley, 2008

20. M. LaPedus, "Toppan rolls 32-nm masks, but can industry afford it?" *EE Times*, 6/13/2008 [Online]. Available: http://www.eetimes.com/electronics-news/4077225/Toppan-rolls-32-nm-masks-but-can-industry-afford-it-

Chapter 2
Radio Systems Overview: Architecture, Performance, and Built-in-Test

In this chapter, we take a look at RF systems from the system to the block level. The overview presents an introduction to the basic communication system, its various architectures and implementations. An architecture best suited for SoC integration is highlighted. Throughout the book, elements will be added to that baseline configuration to enable and demonstrate true built-in self-test and self-calibration capabilities.

Section 2.1 starts with a description of a high-level basic communication system, the heterodyne and homodyne architectures, and their respective considerations. The more suitable architecture for SoC integration is then described. Furthermore, in Sect. 2.2, the important blocks in the radio system are highlighted along with their metrics and critical impairments. The different Built-in-Testing schemes for RF systems are presented in Sect. 2.3.

2.1 Transceiver Architectures

Wireless receivers and transmitters – called *transceivers* when they co-exist in a system – are often symmetrical and parallel in construction but accomplish opposing translations. They both contain elements that perform signal amplification, conditioning, frequency and type (analog versus digital) conversion. Ultimately, they perform the translation from and to wireless high-frequency analog signals and digitally-encoded data bitstreams.

In this section, a basic communication system architecture is first described and then its various configurations briefly highlighted. The different aspects of these configurations are explained and contrasted with their advantages and disadvantages – while highlighting the more suitable setup for integration in RF SoCs. Following that, the important radio system and block metrics are presented along with their impairments. Lastly, we describe the testing of transceivers as it pertains to highly integrated SoCs with primary emphasis on built-in-testability.

S. Bou-Sleiman and M. Ismail, *Built-in-Self-Test and Digital Self-Calibration for RF SoCs*, SpringerBriefs in Electrical and Computer Engineering, DOI 10.1007/978-1-4419-9548-3_2, © Springer Science+Business Media, LLC 2012

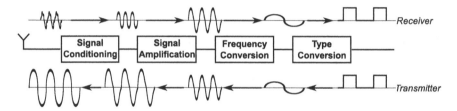

Fig. 2.1 Basic communication system and its constituent blocks

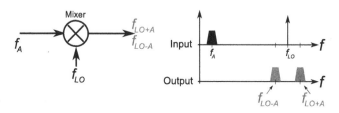

Fig. 2.2 The mixer as a frequency conversion block

2.1.1 Basic Communication System Architecture

Figure 2.1 shows a high-level diagram of a radio system: as shown, receivers operate unidirectionally from left to right, receiving a radio-frequency weak signal from the antenna and propagating that signal along the chain while performing conditioning (filtering), amplification, followed by downconversion of the radio-frequency content to frequencies suitable for digitization. The transmitter is at its simplest the mirror opposite.

The signal conditioning block between the antenna and the amplification block is usually a band pass filter (BPF) that allows through the required frequency while attenuating unwanted out-of-band frequencies. It is very common that this block's functionality is merged into either the antenna (by the inherent nature of its design) or into the amplification stage (e.g. narrowband amplifiers). The amplification blocks in the receiver and the transmitter differ in their design optimizations. In the former, it is important that the amplifier introduces as little noise to the weak signal while amplifying it. In the transmitter, the amplifier is optimized to send a powerful clean signal out through the antenna. As such, the amplifier in the receiver is called a Low Noise Amplifier (LNA) whereas that in the transmitter, a Power Amplifier (PA). Depending on the specifications, multiple stages of amplification might be necessary.

The frequency conversion block comprises of a mixer coupled with a local oscillator (LO), or a cascade of such. Time-domain multiplication of signals, i.e. mixing, results in frequency-domain conversions. The frequency conversion factor is then determined by the oscillator frequency. It is imperative to note that signal mixing produces sideband copies, located at the sum and difference of the respective signals (Fig. 2.2). Proper filtering and amplification are needed to pick out one sideband to propagate the chain. In a receiver, high frequency inputs are downconverted to low

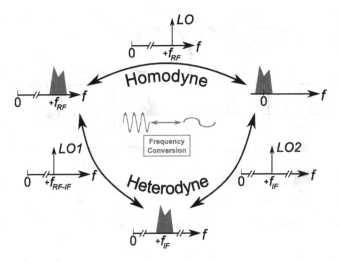

Fig. 2.3 Homoydne and heterodyne frequency conversion

frequencies for easier handling by the following processing blocks. Downconversion is then the result of mixing the signal with the LO frequency followed by appropriate filtering to obtain the lower sideband. On the transmit side, low frequency signals supplied by the processor are upconverted to radio frequencies for transmission.

On either sides of the radio chain, a signal medium and type translation is necessary. On one end, the antenna acts as the air interface to convert between wired and wireless signals and on the other end, a signal type conversion block interfaces the analog and digital domains: Analog-to-Digital converters (ADC) on the receiver side and Digital-to-Analog converters (DAC) on the transmit side.

2.1.2 Heterodyne and Homodyne Configurations

Transceivers can be grouped into two major architectures, heterodyne and homodyne, differing mainly in the number of frequency conversions performed. In essence, the previously discussed high-level system architecture in Fig. 2.1 remains the same with the exception of whether the frequency translation is done in a single shot or through multiple conversions, as depicted in Fig. 2.3.

Heterodyne systems use the latter, and most commonly in two steps. These frequency conversions shift the main signal to an intermediate frequency (IF) first, and then to the desired final frequency, whether it being RF towards the antenna or low-frequency towards the processor. This segmentation allows for more system adaptability and gives the heterodyne systems superior selectivity and sensitivity. However, these advantages come at the expense of bulkier, more complex, and higher power-consumption circuits. Apart from the area and design overhead for two mixers

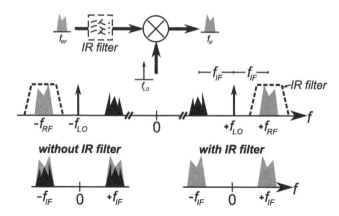

Fig. 2.4 The image problem and the need for image rejection

and LOs, multiple mixing operations demand stringent filtering requirements. For one, band-pass filters operating around IF are needed in-between the two stages for channel selection. However, of more importance is filtering out the "image".

The LO is usually a sinusoidal tone whose frequency contents are two impulses at symmetrically opposite frequencies around dc – positive and negative (as is the case for real signals). This causes the negative frequency tone to also contribute to the mixing. Therefore, signals located at frequencies symmetrical to the desired signal's frequency with respect to the LO tone also exit the mixer at the same IF. To prevent this unwanted band (i.e. "image") from overlapping and corrupting the intended signal, image reject (IR) filters are required. Figure 2.4 depicts the image problem with a desired band at RF frequencies getting corrupted by an image signal when downconverted with and without an IR filter. High quality factor IR filters are often bulky and cannot be integrated in silicon, thus requiring additional routing of high-frequency signals off-chip and back. Therefore, for integrated solutions and System-on-Chip applications, a heterodyne receiver is seldom used.

In contrast, the single-shot conversion scheme is called the homodyne architecture. This architecture was originally developed to overcome the image problems in hetero-dyne systems through direct conversion to – and from – dc or baseband (BB), lending them the name *zero-IF*. This relieves the filtering requirements and results in less components thus making this type of architecture very suitable for SoC integration. However, homodyne architectures inherit their own set of drawbacks. For one, they require very stable and accurate frequency translation, with LO frequencies as high as the target frequency. Moreover, they also suffer from some problematic system and circuit issues that manifest themselves at dc and low frequencies. Nonidealities in the analog circuitry and even self-mixing of the LO signal due to poor mixer isolation result in dc-offsets and low-frequency harmonics that eventually corrupt the desired signal itself. These effects are highlighted in the upcoming sections.

Moreover, low-frequency noise such as device flicker noise also significantly degrades the target signal. Recent advances in data coding and complex modulation

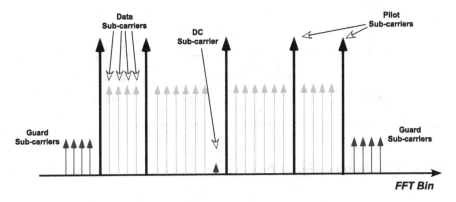

Fig. 2.5 An example signal band with dc-free encoding

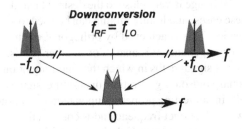

Fig. 2.6 Image problem in direct-conversion architectures

schemes allow for dc-free signals that can be received without distortion due to self-mixing and other dc-offsets. For example, a typical WiMAX and LTE baseband (Fig. 2.5) does not use a dc sub-carrier therefore facilitating dc-offset removal without affecting the received signal.

An intermediary architecture is the low-IF topology. It is a homodyne architecture in construction, but conversion is done to and from a low frequency IF. The dc-offset, LO feed through, noise and various nonlinearities are then of less concern as they do not affect the signal at low IF. On the other hand, direct digitization of the low-IF signal demands high sampling rate ADCs in the receiver and DACs in the transmitter. Also, the image folding still exists at low-IF but its rejection can be migrated to the digital domain.

Even though direct conversion promises image free modulation and demodulation, technically this is not completely true. In zero-IF systems, an image still exists: the signal itself. Apart from amplitude modulated (AM) signals, the upper and lower side lobes of a certain band of interest are not necessarily symmetrical, as is the case for frequency and phase modulation. Therefore, upon conversion to dc, the desired band will overlap with its mirrored self, i.e. the distinct upper and lower side lobes now overlap and corrupt each other (Fig. 2.6). The same phenomenon happens at the second and final conversion of a heterodyne system. To overcome this limitation, a phase and frequency decoupling mechanism is needed – hence enters quadrature mixing.

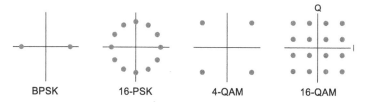

Fig. 2.7 Constellation diagrams for various complex modulation schemes

2.1.3 Quadrature Signal Processing

In zero-IF as well as in the last stage of a heterodyne, the mirror-copy image frequency overlaps with the target frequency band and cannot be solved with filtering. In low-IF scenarios, the image is very close to the desired band thus requiring a very sharp IR filter. In these cases, quadrature mixing can be used to separate the positive and negative frequencies. This is achieved by orthogonal mixing, with in-phase (I) and out-of-phase (Q) components, 90° apart – or in other words, a cosine and sine. This creates two parallel signal paths in which the incoming or outgoing signals are mixed with orthogonal sinusoids generating quadrature signals with phase shift. The phase shift results in interesting properties upon recombination where the image adds destructively while the target frequency adds constructively.

Also an increasing number of today's digital communication standards rely on phase modulation and therefore have quadrature aspects as the basis of their schemes. In these standards, data is encoded and decoded as a complex signal with the in-phase and out-of-phase components as the real and imaginary parts, respectively. This forms an IQ space resembling a two-dimensional Cartesian space commonly referred to as a signal constellation plot or diagram. Figure 2.7 shows the ideal locations of several such schemes where data can take on one of these symbols differentiated by phase shifts (Phase Shift Keying, PSK) or combined amplitude and phase shifts (Quadrature Amplitude Modulation, QAM). For high data rate applications, very dense constellation spaces are needed to encode and decode an increasing number of bits per symbol, for example 64-QAM.

As an image rejection technique, quadrature processing is a very attractive option. However, sufficient image rejection is only guaranteed when the two paths are matched. Any mismatch in the dual paths will result in incomplete cancelation of the image. In the case of digital information carried on the RF carrier, incorrect reception of phase and amplitude might result in interpreting an erroneous symbol, i.e. the real symbol constellation becomes distorted due to the I and Q images. The two mismatch mechanisms here are mainly due to gain variations (amplitude mismatch) or slightly out of quadrature LO (phase mismatch). Appropriate sideband rejection is hence a function of how matched the two paths are, for which a figure of merit is the Image Rejection Ratio (IRR). In the next section describing RF impediments, it will be shown how minor mismatches in either amplitude or phase rapidly deteriorate the IRR.

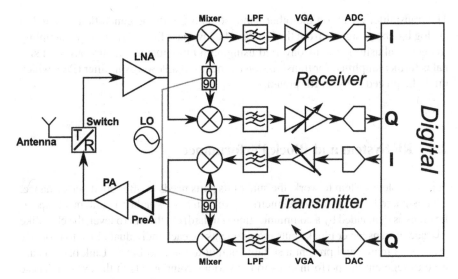

Fig. 2.8 Direct conversion transceiver architecture

2.1.4 Transceiver Architecture for Multi-Band Multi-Standard SoCs

For integrated SoCs with multi-standard capabilities, a flexible transceiver architecture is highly desired. The most appropriate selection would then be the direct conversion architecture – zero- and low-IF: good performance from state-of-the-art LO (frequency synthesizers) and IQ modulators, fewer filters (better wide-band capabilities) and less power, less mixing product spurs, and for the case of newer standards, no active subcarriers at dc.

Depending on the communication system, a transmitter and receiver can function at the same time, given they operate at different frequencies, or alternatively use the same frequency but alternate in operation. These two scenarios are called frequency or time division duplexing, FDD and TDD, respectively. This allows both to share the same antenna while placing a routing block, usually a transmit/receive switch (time-domain splitting) or duplexer (frequency-domain splitting) usually as discrete off-chip components. A typical TDD direct conversion transceiver is shown in Fig. 2.8. TDD systems require only a single LO as the transmitter and receiver use the same frequency band but are duplexed in the time-domain through an RF-switch to and from the antenna. FDD systems on the other hand use different frequencies for the transmit and receive and therefore require two LO signals. In Fig. 2.8, the receiver is composed of a Low-Noise Amplifier (LNA) amplifying the signal with little added noise before being down-converted through two mixers driven by quadrature LO signals. The baseband signal can then be filtered using low pass filters before being amplified by Variable Gain Amplifiers (VGAs) to appropriate ADC levels.

The transmitter chain, on the other hand, starts off with the generation of I and Q analog baseband signals from the DACs to be low-pass filtered (to remove sampling frequency aliasing) and upconverted using an IQ modulator. The upconverted signal is further amplified for transmission by means of a Power Amplifier (PA), which might be preceded by an RF preamp.

2.2   RF System and Block Performance

For a complete system to work, the sum of its parts needs to offer good performance. On the macro level, system level metrics give insight to whether a radio meets specifications as demanded by a communication standard. On the micro level, the chainlike cascade of transceiver blocks puts requirements on each individual block to properly process its inputs and present a suitable output to the next block. Link budgets are used to segment the performances to the various components in the system. These components have their own set of important metrics and have to meet these performance goals for the system to maintain the required overall performance. In the following, we highlight the most important metrics on the system and component levels and trace their dependence on circuit impairments.

2.2.1   System Metrics

Wireless standards often specify certain high-level metrics to ensure transmission quality. Looking at an end-to-end wireless system, it makes sense that the ultimate goal is the transmission and reception of error-free data. Therefore, system metrics such as the bit error rate and error vector magnitude are used extensively to test for standard compliance.

2.2.1.1   BER and EVM

The Bit Error Rate (BER) is a definitive test of the performance of a communication channel. It is defined as the number of bits received in error divided by the total number of bits transmitted during a unit time. As such, BER can represent the end-to-end system performance, from modulation, transmission, propagation, reception and demodulation. Depending on the type of modulation, BER can behave differently with different values of the Signal-to-Noise Ratio (SNR) – the latter being the power of the signal in relation to the system total noise. In general, an increase in SNR causes a decrease in BER as signals are clearly distinguishable from noise and correctly interpreted. Figure 2.9 shows BER versus SNR for a number of modulation schemes showing the quick roll-off with respect to SNR.

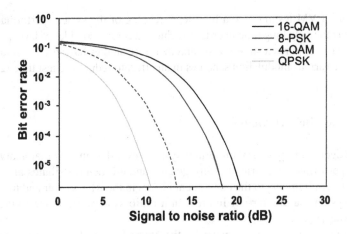

Fig. 2.9 BER versus SNR for various modulation schemes

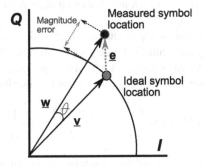

Fig. 2.10 IQ plane with ideal and measured symbol locations

Another metric is the Error Vector Magnitude (EVM). As the digital data is transmitted or received, the actual symbol location, as mapped on an IQ constellation, might differ from the ideal location. EVM represents a measure of this discrepancy and hence provides an indication of modulator/demodulator performance. EVM provides a compacted reading of many parameters that affect the individual blocks of the system, such as poor IRR, phase noise, and carrier leakage (all of which are described in the coming section). The error is computed as the difference vector between the measured and ideal symbol. A graphical representation is shown in Fig. 2.10, where $\underline{v}$ is the vector denoting the ideal symbol location, w the measured symbol location, θ the phase error, $|w|-|v|$ is the magnitude error, and e is the error vector. The error vector magnitude is then defined as $|e|/|v|$, and sometimes referred in percentage by dividing it by $|v|$. The root-mean-square (RMS) EVM and phase error are then used to determine the EVM measurement over a window of several demodulated symbols.

To remove the dependence on system gain distribution, EVM is normalized by $|\underline{v}|$, which is expressed as a percentage, or as a root-mean-squared over a measurement window.

BER and EVM testing give a high level reading of the system performance but both take a relatively long time in order to achieve a certain confidence level [1]. EVM testing usually requires a shorter time and therefore studies have tried to relate BER to EVM for various modulation schemes in an effort to reduce the test time [2].

2.2.1.2 Link Budget Analysis

A link budget study is a very important system level design step that analyzes the required performance metrics for any given standard then distributes and divides the system requirements to the constituent blocks in the RF receiver and transmitter chains. System parameters of interest in a radio system are noise figure, gain, and nonlinearity.

An important performance metric is the Noise Factor (F) which measures the Signal-to-Noise ratio (SNR) degradation from the input of the receiver to its output. The Noise Factor is more often than not described in its logarithmic equivalent the Noise Figure $(NF) = 10\log(F)$.

Each block within a transceiver is characterized with its individual noise factor F_i (or noise figure, NF_i) and gain G_i. Since a typical receiver or transmitter is composed of cascaded blocks, the total gain is simply the product of the individual gains $(G_T = G_1 G_2 ... G_n)$ however the total noise factor can be described as follows:

$$F_T = 1 + (F_1 - 1) + \frac{(F_2 - 1)}{G_1} + \frac{(F_3 - 1)}{G_1 G_2} + ... + \frac{(F_n - 1)}{G_1 ... G_{n-1}} \qquad (2.1)$$

The sensitivity of a receiver, or the minimum power of a detectable signal, is then defined as:

$$Sens_{RX} = kT + 10\log(B) + SNR_{min} + NF_T \qquad (2.2)$$

where kT represents the power spectral density of thermal noise, B the noise bandwidth, SNR_{min} the minimum required by the ADC to supply a sufficient BER, and NF_T the total noise figure of the receiver chain.

As sensitivity and hence the weakest desired detectable signal demand minimum SNR and NF requirements, gain cannot be driven to be arbitrarily large as we are hit with the receiver blocks' degraded linearity performance and the ADC limitations (dynamic range – or maximum acceptable signal). Therefore, unlike what is expected, we note an increase in BER with a very large received signal strength – much like when the received signal is weak. Therefore, radio systems are not only bound on the lower end by the noise floor but also on the upper end by the inherent nonlinearities of the circuits.

System blocks are non-ideal and therefore do not perform always linearly – keeping in mind that some blocks' functionality is based on non-linearity (mixers for example). However, if we look at receivers, nonlinearities bring rise to gain compression, intermodulation distortion, desensitization, and cross-modulation. These undesired

Fig. 2.11 Two-tone intermodulation spectrum

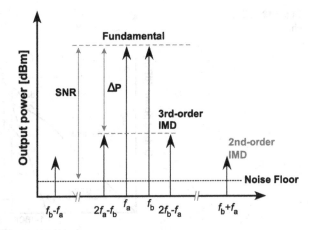

effects could manifest when there is enough signal power either in the desired signal itself or possibly in a close-by interferer (or blocker). The 1 dB compression point, P_{1dB}, represents the input power that experiences a 1 dB decrease from the expected linear gain. When two closely spaced tones enter a non-linear system – say a desired signal and an interferer or possibly two interferers close to the desired band – mixing is bound to happen and intermodulation products appear in the spectrum, as depicted in Fig. 2.11. Second order intermodulation products fall far from the band of interest, at the sum and difference frequencies. While the former can be easily filtered, the latter is problematic in the case of zero-IF receivers as it falls close to dc. Therefore, second order intermodulation is problematic for mixers. Third order intermodulations are the hardest to get rid of as they fall extremely close to the two tones and within the band of interest, in case of an LNA for example. Third order intermodulation is tested by applying two closely-spaced and equal power tones. The intercept point at which the third order intermodulations, that increase cubically with input power, equal the main tones (that increase linearly with input power) is called *IP3*, and consequently the input power at that point is called *IIP3*, as shown in Fig. 2.12. As with the total noise factor, the cascade of receiver blocks results in a total *IIP3* measure given by:

$$\frac{1}{IIP3_T^2} = \frac{1}{IIP3_1^2} + \frac{G_1}{IIP3_2^2} + \frac{G_1 G_2}{IIP3_3^2} + \dots \qquad (2.3)$$

The higher the gain at the first blocks the more stringent the linearity requirements for the preceding blocks. Additionally, care should be taken to never exceed the ADC dynamic range. Therefore, the receiver chain should be able to (1) provide a large enough gain to amplify weak signals to meet a minimum SNR by the ADC and (2) make sure that a strong received signal is not amplified beyond the following blocks' input range or that of the ADC. These two requirements might not result in a single converging link budget; one way to tackle this is by using a dual gain frontend with either the LNA/PA or mixer having two gain modes: A high gain

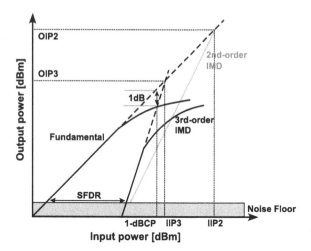

Fig. 2.12 Linearity characteristics

mode to satisfy the sensitivity and noise requirements and a low gain mode to satisfy linearity constraints. Moreover, Automatic Gain Control (AGC) is also employed to cover a range of input powers to satisfy dynamic range requirements, or more commonly the spurious free dynamic range (SFDR). The SFDR is defined as the maximum input level where intermodulations do not exceed the noise floor in relation to the minimum detectable input power.

A careful link budget analysis is therefore necessary to realize a system that offers conformance to the standard's specification while allotting the required performance specifications for the individual building blocks and components.

2.2.2 Component Metrics

The RF blocks that constitute the receiver or transmitter chains provide the required signal processing for establishing communication. Each of these blocks performs a certain task and suffers from a number of impairments limiting its effectiveness and performance. Receiver chains are more challenging, as they have to deal with the reception and demodulation of very weak signals. In this section, we briefly highlight the metrics and impairments of each of the blocks in the RF chain.

2.2.2.1 Signal Amplifiers: LNA and PA

The Low Noise Amplifier, as its name suggests, has to provide sufficient gain at the minimum amount of added noise. Hence its gain, G, and noise figure, NF, are of primary importance. Also, as this block interfaces to the antenna through a T/R switch or duplexer it has to present proper impedance matching for maximum power transfer of the weak received signal. The PA's metrics are similar to those of an LNA

Fig. 2.13 Ideal and nonlinear amplifier effects

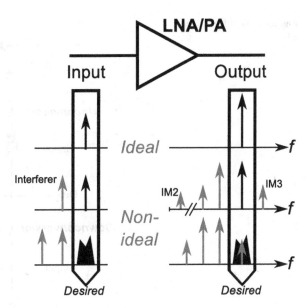

but are more steered to providing sufficient power, good efficiency, and matching to the antenna.

Both LNA and PAs have to provide superior linearity. Designers strive to obtain the highest possible compression points and IIP3s to reduce the emergence of harmonic and intermodulation products. These spectral elements, if not treated appropriately with filtering or otherwise, can interfere with other frequency bands off-chip or on-chip. Off-chip, the PA should prevent spectral regrowth to adjacent channels that can be occupied by other users. Standards usually specify a certain transmit mask that products must adhere to. On-chip, the LNA's output might contain third order inter-modulation products as well as the main tone, with both delivered to the mixer for down conversion. Figure 2.13 shows an ideal and non-ideal amplification, with the latter depicted as intermodulation distortion due to out-of-band interferers.

Typically, one-tone and two-tone tests are performed on LNAs to quantify their gain, noise, and linearity parameters. Power detectors can be used at the inputs and outputs of these amplifiers to measure their signals and their contents and extract the important metrics.

2.2.2.2   Mixer

As described previously, the mixer is a three-port device with two inputs and one output. The signal of interest comes in from one port and exits the other. A local oscillator provides the other input port with a tone for frequency translation. The mixer ideally performs a sum and difference conversion in the frequency

Fig. 2.14 Nonideal mixer effects

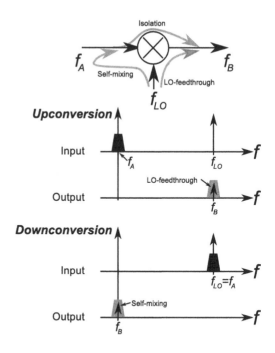

domain that is coupled with a gain (active mixers) or loss (passive mixers). Therefore, an important metric is the conversion gain as well as linearity (second order intermodulation is critical here) and NF. Also, the mixer's conversion gain is a function of the input LO power and the two must be carefully designed in conjunction, especially in IQ-type systems.

Non-ideal isolation between the input ports results in self-mixing, feed through, and in-band intermodulation products (Fig. 2.14). A powerful LO can seep into the second input and mix with itself – its down conversion will become an undesirable dc offset. On the other hand, low isolation can also result in LO power to appear at the output close or coinciding with an upconverted signal in a transmitter.

2.2.2.3   Local Oscillator

The local oscillator is usually implemented as a frequency synthesizer, generating precisely controlled frequencies based on phase locking concepts. A Phase Locked Loop (PLL) uses feedback techniques to lock a high-frequency free running oscillator to a clean and steady low-frequency source.

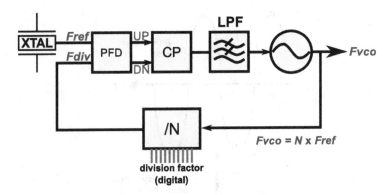

Fig. 2.15 PLL block diagram

PLLs are complex systems comprised of multiple types of circuits, spanning the digital, analog, and RF domains. In general, the inputs to the PLL are a clean low-frequency reference frequency (F_{REF}), usually supplied by a crystal, and a multiplication factor (N, a division value in reality) and the output is a stabilized high frequency multiple of the reference ($N \times F_{REF}$). Figure 2.15 shows a block diagram of a PLL. The negative feedback works on reducing the error between the divided voltage controlled oscillator (VCO) frequency and a stable crystal reference. The comparison block is the phase frequency detector (PFD) which outputs pulse-width-modulated error signals signifying the difference between the two inputs. The error/correction pulses, UP and DN, indicate a lead or lag between these inputs. The PWM signals are supplied to a charge pump (CP) that injects or removes a proportional amount of charges (current) into a smoothing filter. The latter converts the charge input to a voltage that eventually modulates the VCO.

PLLs as frequency synthesizers need to provide programmable frequency shifts to accommodate the different channel requirements dictated by the communication standards. Important synthesizer metrics are the center frequency, lock range, settling time, and phase error. The center frequency needs to be free of frequency and phase offsets to ensure proper mixing and signal transmission/reception. The synthesizer has to be able to lock to all the frequencies of relevance and when instructed to change frequencies, do so in reasonable time (settling time).

A real LO signal is in reality not a clean sinusoid and does not resemble a single tone in the frequency domain, due to phase noise. Phase noise is considered one of the most important parameters in a frequency synthesizer. It is the frequency domain representation of jitter and manifests itself in frequency deviations from the ideal frequency (Fig. 2.16). Single-side-band (*SSB*) phase noise is measured in dBc/Hz and gives the noise power (normalized to 1 Hz bandwidth with respect to the carrier) at given frequency offsets from the carrier. Phase noise is also integrated over a frequency band to provide a single value, either expressed in dBc or in degrees as phase error. Root-mean-squared phase error can then be easily transformed to timing jitter.

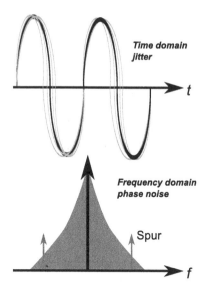

Fig. 2.16 Jitter and phase noise

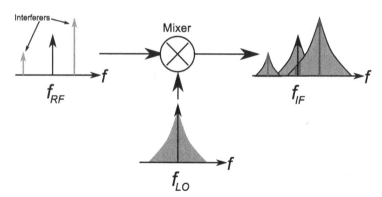

Fig. 2.17 Effects of phase noise on mixing

Each component of the PLL contributes in one way or another to the phase noise of the output signal: some blocks contribute noise close to the synthesized signal while others' noises dominate at higher offsets. Also, apart from random noise, high-powered discrete noise tones can appear in the output, called spurs. A common spur in PLLs is the reference spur which appears at F_{REF} offsets from the center tone. It is mainly caused by the PLLs' periodic correction of some of its blocks' non-idealities (loop filter leakage and charge pump current mismatch).

Since the PLL is used as a local oscillator to perform frequency up- or down-conversion, the non-ideal tone with "*skirts*" endangers a correct transmission or reception of a signal. This can be seen in Fig. 2.17: the noisy local oscillator causes one of the weak RF tones to be distorted by the dominant phase noise of the stronger tone. Therefore, the various communication standards impose a *mask* on the phase noise spectrum of local oscillators.

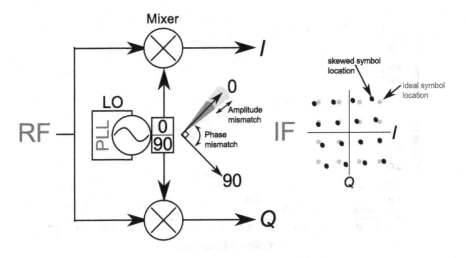

Fig. 2.18   IQ demodulator: effects of amplitude and phase mismatch

2.2.2.4   IQ Modulators and Demodulators

IQ modulators and demodulators are simply a special construction of LOs and mixers. The principle of operation was described previously as a method for image rejection. Since digital data is encoded and decoded with quadrature content on the digital level, it needs to be transmitted and received without the mirror images corrupting each other. Quadrature LO signals need to be generated and fed to two mixers, one on each path, as is shown in Fig. 2.18. It is important here that the matching between the I and Q channels be superior as minor mismatches will reduce the suppression of the mirror images and corrupt the intended symbols.

The important metrics here are amplitude and phase mismatch. The 0° and 90° signals have to be of the same amplitude and in perfect quadrature. Also, differences in the two mixers' gains contribute to the mismatch. If the amplitude and phases of the paths are not matched, the image rejection ratio, which represents how much of the undesired image has been suppressed, will be adversely affected. The IRR as a function of amplitude and phase mismatch is given by:

$$IRR_{dB} = 10\log\left[\frac{\alpha^2 - 2\alpha\cos\varphi + 1}{\alpha^2 + 2\alpha\cos\phi + 1}\right] \tag{2.4}$$

where α is the amplitude imbalance (expressed as a ratio) and φ is the angle mismatch from perfect quadrature between the two paths. A value greater than 60 dB is often desired but slight imbalances cause a disproportionate drop in that value. Figure 2.19 shows how sensitive the IRR is to small fluctuations in the amplitude and phase matching.

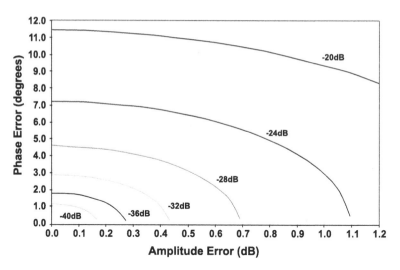

Fig. 2.19 Phase and amplitude imbalance effects on image rejection

2.3   Integrated Radio and System-on-Chip Testing

The previously discussed system and block metrics need to be verified to ensure proper functionality. With integrated RF SoCs, both digital and RF blocks coexist on chip and testing becomes more involved. On one hand, digital testing has reached a mature stage and its techniques have adopted well-known and widely applicable fault models [3]. On the other hand, testing for RF and analog, in general, still relies on checking for conformity to a set of design specifications. However, observing these performance parameters is quite a challenge as RF systems, especially highly-integrated transceivers, offer little accessibility to individual blocks. Limited accessibility limits the detectability of faults and restricts observability. This is especially true for the traditional RF testing methods where powerful mixed-signal Automatic Test Equipment (ATE) and benchtop laboratory setups – such as rack-and-stack – require a sizeable investment in terms of actual cost to obtain and maintain, difficulty of interfacing to the internal nodes, and long test times.

A promising technique borrowed from digital design and only recently applied to RF is Built-in-Self-Test, or BiST. On-chip BiST provides an opportunity to control inputs and observe outputs of individual blocks as well as signal chains. This is enabled by the insertion of simple measurement circuits at critical internal nodes. These sensors can then provide readings indicative of signal properties of interest at inputs, outputs, or even the insides of individual on-chip blocks. Moreover, an extension to mere sensing is also the inclusion of on-chip test stimulus generators for a truly internal test cycle. BiST techniques can therefore be cost-effective alternatives to, albeit not necessarily more accurate than, traditional testing equipment for complex integrated RF SoC.

Although neither ATE, Rack-and-Stack nor BiST techniques are used exclusively, only the latter promises beyond-post-production test portability. This is a very important enabling technique to ensure the viability of microchips with an extended set of operating conditions including process, voltage, and temperature changes (PVT). This gains more importance when we discuss self-calibration and self-healing techniques, for which BiST is a required precursor.

Having successful and efficient on-chip testing paradigms not only increases the probability of detecting faulty blocks but also allows for post-silicon quality measures, of which on-the-fly tuning promises to offer robustness enhancement for heterogeneous systems such as RF SoCs. It has become quite evident with the increased integration that manufacturing testing is not enough for validating a complex system. In-field and operational variability needs to be taken into account; and to lump all these test cases at the production testing stage is a daunting task. Therefore, there is a need for increasing detectability and allowing measurement paradigms to be integrated into the circuitry rather than being outside.

2.3.1 Built-in-Test Techniques

Migrating the testing on-chip, different techniques have been proposed and used, either stand-alone or as enhancement to traditional tests. Several low-cost testing methods exist such as *loopback testing*, *alternate testing*, and *digitally-assisted testing*. These methods are not mutually exclusive and can be used together to achieve the desired BiST functionality.

Loopback testing is among the most well known methods, especially for RF systems employing both a transmitter and receiver. Its advantages include low cost, extremely low hardware overhead (if any), and simple (usually a single-metric) testing. Cost and hardware overhead are very low because of the high levels of component re-use. The digital baseband and RF sections of the system are tested together by forming a loop from the transmitter to the receiver. A digital bit-stream is generated in the transmitter's baseband, transformed and upconverted to RF, and then coupled to the receiver [4]. The coupling circuitry is usually the only additional overhead. This coupling can be achieved internally (on-chip) or externally (on-board). However, depending on the type of duplexing the system uses, the hardware needed might slightly change. For example, at the minimum, a simple controllable attenuator can be used between the transmitter's pre-amp and the receiver's LNA in a zero-IF TDD system (Fig. 2.20). In contrast, systems whose transmitters and receivers have a frequency offset, such as in FDD systems, proper frequency translation is required prior to coupling. Narrowband systems might be able to perform the translation in the transmitter baseband; broadband systems would need a dedicated mixer with the attenuator [5, 6].

Loopback testing often offers a single-metric measurement, usually the BER (or EVM). BER measurements allow for effective fault detection as they can be used to

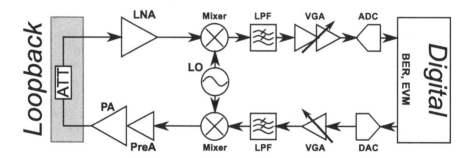

Fig. 2.20 Loopback testing configuration

trace the degradation due to noise and gain impairments in the system. While this is an attractive feature for pass/fail testing, a major downside to loopback testing is fault-masking. Since testing is done end-to-end with the entire system treated as a single block, there is no information on which component is causing failure in the system. This makes fault-localization very difficult especially in complex RF and analog chains. Some methods do exist to alleviate these issues, such as path sensitization and internal node monitoring. In the first, specially crafted test pattern can be transmitted, looped back, and upon reception analyzed for distinct signatures. Faults, if they exist, can be attributed to specific blocks based on the received signal's signature. This technique, however, requires some behavioral modeling effort to quantify appropriate test patterns and their responses. Another method is to augment fault observability by actually monitoring internal nodes in the system, at the expense of additional hardware overhead. On-chip sensors can be inserted to provide more signal information along the loopback chain [7].

Alternate testing is another on-chip low cost testing method that is geared towards characterizing individual components rather than end-to-end specification checking. Alternate tests do not attempt to sense or measure a certain circuit parameter directly but opt to get a reading that can be explained by a set of circuit parameters. Therefore, it presents an attractive alternative for decreasing test time with the possibility of extracting multiple circuit parameters in a single test, and predicting the specifications accordingly. In essence, alternate testing is a correlation testing methodology in which several spaces are mapped to each other: parameter, signature, and specification spaces (Fig. 2.21) [8]. Parameter spaces can be constructed from known circuit and process variations. Then suitable test stimuli are created to expose appropriate signatures in which several distinct circuit parameters are distinguishable. The response of the circuit to the test stimulus constructs the signature space. Both the parameter and corresponding signature spaces can be mapped to a specification space. The acceptance region (pass/fail zones) needs to be carefully determined and mapped. This may require effort on the side of modeling and statistical simulations (Monte Carlo) or even actual measurements on test silicon runs with enough samples [9].

A main objective of alternate testing is the use of inexpensive test generation and signature detection. The alternate testing premise can be contrasted with the

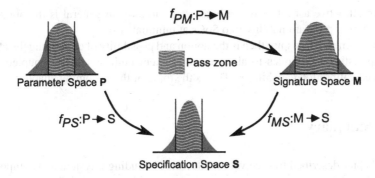

Fig. 2.21 Alternate testing: parameter, signature, and specification spaces and their mapping [9]

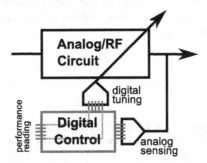

Fig. 2.22 Digitally-assisted analog/RF circuit

previously discussed path sensitization in loopback testing: Path sensitization is applied to systems with many cascaded blocks in order to enable extraction of a specific component parameter whereas alternate testing predicts multiple block parameters from tests on a single component. A downside of alternate testing is the need for solid mappings between the different spaces – process shifts that alter the statistics and correlations require retraining of the models and the corresponding spaces to retain the correct acceptance zones [9].

A testing approach emerging with the proliferation of complex transceiver chips is *digitally-assisted testing*. This testing paradigm builds on the design of digitally-assisted circuits (Fig. 2.22). A number of transceiver blocks now comprise not only of analog but also digital parts with some of the analog tasks migrating to the digital domain. These circuits' digital parts are used to internally monitor performance and tune the analog parts according to predetermined optimizations and calibrations, all in a closed loop manner.

With the built-in capability to digitally monitor and tune the analog performance, specifications can therefore be extracted by observing or reading the digital state. For example, many analog biasing voltages are supplied by DACs in an effort to overcome PVT variations with some degree of programmability; therefore, a reading of the digital word in the DAC can be indicative of the analog block's performance.

A more attractive aspect of digitally-assisted circuitry in general is the ability to embed tuning programmability into SoCs at minimal cost.

What comes to transpire then is the notion and possibility of combining the above on-chip testing techniques to allow more efficient built-in-self-test coupled with enhanced calibration capabilities. This is the topic of the upcoming chapters.

2.4   Summary

This chapter described transceiver architectures including a system and component overview. The most relevant metrics were presented and a brief discussion on the component non-idealities and their effects on proper system functionality. Built-in testing techniques are also described. The material presented in this chapter form the backbone of the suggested RF built-in-self-test and digital self-calibration to be discussed in the later chapters.

References

1. S. Bhattacharya, R. Senguttuvan, A. Chatterjee, "Production test technique for measuring BER of ultra-wideband (UWB) devices," *IEEE Transactions on Microwave Theory and Techniques*, vol.53, no.11, pp. 3474–4381, 2005
2. M. Lin, Q. Zhang, Q. Xu, "EVM Simulation and it comparison with BER for different types of modulation," *TENCON 2007–2007 IEEE Region 10 Conference*, pp.1–4, 2007
3. L. Wang, C. Stroud, N. Touba, *System-on-Chip Test Architectures: Nanometer Design for Testability*. Elsevier, 2008
4. M. Onabajo, M. Silva-Martinez, J. Fernandez, E. Sanchez-Sinencio, "An on-chip loopback block for RF transceiver built-in test," *IEEE Transactions on Circuits and Systems II: Express Briefs*, vol.56, no.6, pp.444–448, 2009
5. J.J. Dabrowski, R.M. Ramzan, "Built-in Loopback Test for IC RF Transceivers," *IEEE Transactions on Very Large Scale Integration (VLSI) Systems*, vol.18, no.6, pp.933–946, June 2010
6. E. Garcia-Moreno, K. Suenaga, R. Picos, S. Bota, M. Roca, E. Isern, "Compact Frequency Offset Circuit for Testing IC RF Transceivers," *Solid-State and Integrated Circuit Technology, 2006. 8th International Conference on ICSICT '06*, pp.2125–2128, Oct. 2006
7. S. Bhattacharya, A. Chatterjee, "A built-in loopback test methodology for RF transceiver circuits using embedded sensor circuits," *Test Symposium, 2004. 13th Asian*, pp. 68–73, Nov. 2004
8. P.N. Variyam, S. Cherubal, A. Chatterjee, "Prediction of analog performance parameters using fast transient testing," *IEEE Transactions on Computer-Aided Design of Integrated Circuits and Systems*, vol.21, no.3, pp.349–361, Mar 2002
9. R. Voorakaranam, S.S. Akbay, S. Bhattacharya, S. Cherubal, A. Chatterjee, "Signature Testing of Analog and RF Circuits: Algorithms and Methodology," *IEEE Transactions on Circuits and Systems I: Regular Papers*, vol.54, no.5, pp.1018–1031, May 2007

Chapter 3
Efficient Testing for RF SoCs

The implementation of self-test, as discussed in this book, is not geared exclusively for production testing but also for post-production. Post-production, or more specifically, post-deployment testing will provide the platform for the implementation of on-chip adaptive calibration. The goal then is not to pass production testing only but to ensure that a product does not operate at the edges of the pass zone, but always comfortably in its optimal region. This is not only to satisfy silicon yield but also to ease the ever increasing complexity of the design process. Designing in nanometer CMOS is becoming a daunting task with the increase in numbers of corners, heralded by process variability and volatility with temperature and power. Faced with the non-optimality of over-design, circuit designers are looking into implementing more innovative techniques – those of assisted operation, therefore boosting performance of the otherwise non-optimal circuit with a robustness enhancer. The concept of tunable RF circuits, and what techniques are best suited to augment their capabilities in SoCs will be discussed in the next chapter. The challenge then becomes how to tune these circuits optimally, over corners, and over time, meaning while operating in a consumer setting. For optimum tuning, the operating condition of these circuits needs to be assessed, beyond the lab or fab. This is only manageable by putting measurement capabilities into the system, hence self-test.

In this chapter, we lay down the requirements for RF SoC BiST and propose a setup that makes maximal use and re-use of available resources, with little additional circuitry. Then, we introduce a measurement circuit that balances the often conflicting requirements of efficient and accurate measurements, insertion non-invasiveness, wide dynamic range, and broadband operation. The proposed circuit is a modification on RF amplitude/power detectors whose design, characteristics, and three different implementations are described.

S. Bou-Sleiman and M. Ismail, *Built-in-Self-Test and Digital Self-Calibration for RF SoCs*, SpringerBriefs in Electrical and Computer Engineering, DOI 10.1007/978-1-4419-9548-3_3, © Springer Science+Business Media, LLC 2012

3.1   On-Chip Test Migration and Portability

To truly migrate the testing functionality on-chip and enable test setup portability, an efficient self-test paradigm for transceivers and radio SoCs should include the ability to complete the test loop internally. Also, the capability to test one or more blocks in the chain, as well as whole system characterization, should be addressed. The applicable tests and the controllability of the test signals are also of concern, especially that the test signals under consideration here are at radio frequencies and beyond.

Therefore, we opt for an architecture that inherits and includes desirable features from each of the built-in-testing techniques described in the previous chapter. Loopback testing's emphasis on system-level testing (such as BER and EVM) needs to get decoupled to enable specification-based direct and alternate testing of individual blocks. Moreover, as these blocks move to digitally-assisted designs, the luxury of tuning them and extracting more performance information from them becomes quite apparent for characterization purposes and self-calibration possibilities. Here is where we draw the line between regular built-in test (BiT) and built-in-self-test (BiST). The latter's self-sufficiency and non-reliance on any off-chip external test equipment is something only possible with heterogeneous single-chip systems. Therefore we need to highlight the following requirements for an on-chip on-the-fly test and calibration setup:

1. *Test signal generation*: Depending on the type of metric sought, the testing technique will differ. The system should have a somewhat flexible test pattern and signal generation. Given the application in the context of a SoC, several resources can be used to generate such signals. To increase resource re-use and reduce overhead, the presence of a flexible DSP in the digital backend can be leveraged to customize test signals, in line with digital signal synthesis [1]. The test signals can be representative of symbols, such as patterns used for BER and EVM tests, or the digital equivalents of analog tones, for use in component level specification testing. The latter setup combines the DSP with the interfacing transmitter DAC and analog baseband to generate single or multi-tone test sources, at baseband frequencies. A sufficiently wideband analog baseband is useful in such cases to create a wider range of synthesizable signals. These lower frequency signals can then be appropriately modulated to RF and be routed internally.

2. *Test signal routing*: The end-to-end test-bench is analogous to the loopback testing technique whereby a transmitter-receiver coupling, or signal routing, is achieved by switches, a programmable test attenuator, and possibly a mixer for transceivers with frequency offsets (e.g. FDD). Insertion of these loopback blocks should not affect the regular operation of the transceiver, and need to be designed accordingly [1, 2]. These blocks should match to the output of the PA and input of the LNA, taking in a relatively large and powerful signal and outputting a controllable signal. Also, it is preferable that they demonstrate superior linearity – putting passive implementations at a slight performance advantage but at higher area cost. Resistive ladder attenuators as well as MOS-based implementations offer good linearity, matching, and range of attenuation [3]. We note here that continuous

attenuation is not really necessary and a finite and discrete set of attenuation levels suffices. A passive mixer in the loopback element is preferred as it has high linearity and inherent attenuation (loss) – with the possibility of having a programmable conversion gain. Both the attenuator and mixer can be programmed to vary their gain and therefore create discrete sweep scenarios. The range of attenuation should encompass all testing needs. For example, the lower limit of attenuation should still allow for tests that require large signals, such as compression tests, to be performed between transmitter and receiver. Therefore, it is very important to analyze these critical loopback elements [4].

3. *Internal node accessibility*: Apart from the loopback element that establishes the path between the transmitter and receiver and enables variable attenuation levels, RF components in the middle of the chains cannot be directly accessed for characterization and individual testing. Bypassing techniques can be used to route signals to a specific block while also turning off bypassed circuits for better signal integrity. Low insertion loss switches are needed with very high isolation in the off state to ensure that signals pass through with minimal degradation. To decrease the insertion loss, bigger switches are needed however this increases their capacitive contributions at high frequencies. The size of the switches should be taken into consideration when designing the rest of the circuits [5].

4. *Internal node visibility*: One of the downsides of loopback testing is the lack of information on the internal nodes of the end-to-end setup. This has been alleviated by the inclusion of small on-chip RF detectors that attach to these nodes and provide an easily readable measure of the RF signal. A common output of these detectors is a dc value corresponding to either peak amplitude or RMS power. We will leave the more extensive description to the next section that presents existing RF detectors and proposes a more suitable implementation for RF SoC BiST. However, we mention here a few notes pertaining to this block. RF detectors, much like the test attenuator and switches, are additional circuits and as such their presence needs to be as non-invasive to the system as possible. This places certain requirements on their design. Multiple detectors can be placed along the transmitter and receiver chains, with their outputs forming a low frequency (mostly dc) bus that can be easily digitized for analysis.

5. *Test result analysis*: The powerful digital backend can also be used to perform test results analysis. Test results can be derived from two sources: the primary digital lane and the auxiliary dc lane. The digital lane is simply the digitized I and Q channels on the receiver. This is in essence the regular return test path for a loopback BiT where two channels interface to the digital processor through their respective ADCs – which is also the case under normal transceiver operation. On the other hand, the auxiliary dc lane is a test-only path that holds the test data from the RF detectors. These dc values can be digitized using the system ADCs and their readings used to extract performance parameters.

6. *Overhead*: An important recurring point in the previous discussion is the stress on lowering the overhead – in area, power, and test time – of the new embedded capabilities. Regarding area, a suitable architecture will make maximal use of existing hardware and only require a few additional circuits (attenuator, switches,

offset mixer, and detectors) – normally well below 10% of the total area [6]. The major benefit here is resource re-use, and more importantly the powerful digital hardware which when coupled with a malleable software layer (e.g. algorithms) can offer high levels of flexibility and customization. As testing is only performed intermittently, power and time overhead can be easily optimized. When not testing, the additional circuitry can be turned off, e.g. power gating. Moreover, testing can be scheduled by the system during down time.

The next section presents a complete transceiver with the required modifications for enabling RF BiST and BiSC.

3.2 A BiST-Ready RF SoC

Figure 3.1 shows the modification of the transceiver architecture previously presented in Chap. 2. The changes are done to satisfy the requirements set forth in the previous section. RF signals can be created in the baseband of the transmitter and upconverted by the mixers. A router-type loopback element interfaces the transmitter to the receiver. Switches accomplish routing of the test signals and their attenuation is further controlled by the programmable attenuator. Bypassing the front elements of the transceiver also deactivates them so that they do not affect the signals being routed around them.

The system shown here is a TDD system where a single LO is used for the transmitter and receiver. In the case of an FDD system with two LO blocks, the frequency offset between the receiver and transmitter can be synthesized in the transmitter baseband or appended in a loopback mixer. All blocks in the system, even the loopback elements, can be monitored by the RF detectors that are scattered around the two chains. The presence of this small detector eliminates fault masking and enables the

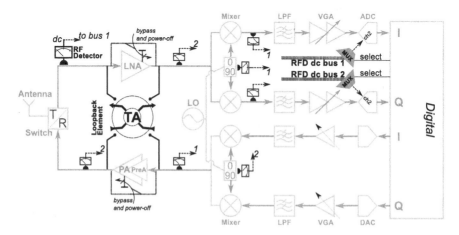

Fig. 3.1 A BiST-ready RF SoC

Table 3.1 Important transceiver parameters to measure

	LNA	Mixer	PLL (LO)	PA
Gain	•	•		
Output power			•	•
Linearity	•	•		•
Input match	•			
I/Q match		•	•	
Noise	•	•		•
Phase noise			•	

testing and monitoring of the test generation circuitry before signals are routed for testing. This ensures test signal integrity and enables signal tracking along the chain. The multiple detectors' outputs form two dc busses going to the receiver ADCs. A main multiplexer toggles the input of the ADCs between normal and test modes, sending the demodulated quadrature signals in the first case and detectors outputs in the second. Successive detectors are assigned to separate ADCs to allow for concurrent input and output readings of a single block. In the case that the required detectors outputs are assigned to the same ADC, then readings can be time interleaved.

Internal to the digital part is the DSP responsible for the testing algorithms. The algorithms specify the test signal creation and control the test circuitry, i.e. switches, attenuator, and detectors. They also perform the analysis to extract the blocks' performance parameters. Table 3.1 highlights the important parameters to test for in an RF SoC. Several of these parameters can be extracted using power and amplitude measurements while others require spectral measurements. The general one-tone and two-tone tests can be generated in baseband for parameter extraction. Compression and intermodulation points can be measured by successive sweeps of these tests. Other parameters can be indirectly deduced from alternate test result processing. Possible measurements in this system will be highlighted later in this chapter.

A more intriguing capability now arises with circuit self-awareness. Based on the results of the tests, the transceiver blocks can be tuned to optimize their performance. As the circuit implementations of the RF blocks makes use of the digitally-assisted methodology, calibration DSP algorithms can make sense of test results to change the state of the block under calibration. BiSC will be discussed in Chap. 5.

Earlier implementations of loopback with increased observability through RF detectors still relied on outside equipment to properly analyze the output signatures of the detectors [6, 7, 11]. True BiST should only use internal components and therefore maximally benefit from the ADC and DSP at the back of the high-frequency transceiver. The measurement, or reading, accuracy is then limited by the internal ADC resolution and the RF detectors' sensitivity. Higher resolution ADCs will enable better discrimination between dc outputs and high sensitivity detectors will offer better distinction between slight changes in RF features. The ADC resolution is primarily set by the system requirements, with ranges in the order of 10–12-bits in state-of-the-art ADCs. Therefore, to improve testing accuracy, the RF detector's sensitivity should be increased.

3.3 RF Amplitude Detectors for RF BiST

This section describes the sensor to be used in the on-chip BiST and BiSC. The RF amplitude detector presented here is capable of broadband operation with wide dynamic range and high sensitivity. The requirements for such a detector for successful embedding into a self-test scheme are listed along with previous detectors in literature and their shortcomings. The most prevalent types of detectors convert high-frequency signals – either sensitive to amplitude or power of a signal – to a corresponding dc voltage. This dc voltage is a digitally-friendly mapping of the RF signal which upon digitization will represent a reading of the signal state. Following in this section, the proposed amplitude detector is presented along with its analysis, design, and circuit implementations.

3.3.1 Detector Requirements

For parameter extraction of the various blocks in the transceiver, multiple detectors need to be inserted along the RF chains. Since the signal levels and properties differ between these blocks, a suitable detector needs to be designed to ensure accurate measurements chain-wide. Moreover, much like the additional test circuitry, the detector should be as non-invasive as possible so as not to affect the normal operation of the block and signals it is monitoring. The design of an appropriate detector for RF SoCs should meet a number of requirements [11]:

1. *Small area and low power:* These are major requirements, as multiple detectors are needed in the loopback chains. A detector should be only a fraction of the main block it is monitoring which limits the type of circuits and device sizes that can be implemented. Power is less of an issue if the detector can be turned off when the system is not in test mode.
2. *Non-invasiveness:* The detector has to connect to the signal path but should be designed not to load it. Therefore, it should be transparent to the system. Designing the detector with a high input impedance can ensure minimal loading, which is critical in the impedance matched blocks like the LNA and PA.
3. *Wide dynamic range:* The dynamic range represents the range of amplitudes that can be sensed. In an effort to have a single detector implementation for the entire system, the designed circuit should withstand widely varying amplitude signals while outputting a correct dc value.
4. *Broadband operation:* The detector's ability to cover a wide range of frequencies will enable its use in multi-standard RF SoCs.
5. *Accurate and sensitive response:* A stable and accurate high-frequency-to-dc conversion will allow for very fine detection of small amplitude changes. A stable detector response ensures that PVT variations will not affect the measurement while a high conversion gain from RF to dc will decrease the minimum detectable amplitude change, given by

$$\min\left(\Delta amp\right) = \frac{V_{fullscale}}{2^{ADCbits} A_{rf-dc}} \tag{3.1}$$

where $V_{fullscale}$ is the ADC fullscale, $ADCbits$ the number of bits of the ADC and A_{rf-dc} the RF-to-dc conversion gain.

The above requirements place guidelines for the design of a suitable detector for inclusion in an RF BiST scheme.

3.3.2 Detector Architectures

CMOS on-chip amplitude detectors have recently gained importance for Built-in-Test applications to enable internal probing of RF transceivers. Several implementations have been published in literature [8–17]. The basic premise is the use of the non-linear properties of MOS transistors to convert the high frequency input signal to a low frequency (and dc) current, which in turn creates a voltage over a load. Single-ended and differential implementations are possible depending on the type of the circuit being monitored. Table 3.2 lists some properties of the CMOS RF and mm-wave detectors in literature and Fig. 3.2 plots their characteristic conversions.

From Fig. 3.2, it can be seen that most detectors have a low conversion gain and some have a narrow amplitude detection range. An architecture that has both a high conversion gain and wide amplitude range is presented in the next section.

3.3.3 Proposed Detector Design

The proposed detector, shown in Fig. 3.3, is a simple single-stage implementation quite similar to some of the previously mentioned implementations [8, 11, 17]. However, its implementation includes modifications to increase the conversion gain and widen the dynamic range. The input of the detector is ac-coupled to allow for separate biasing to the gate of the input device. The output is taken out at the low-pass filter end of the detector. The load is either an active device or simply a resistor – in the actual implementations discussed later either is used. The detector presents a capacitive load to the path it is monitoring; hence, its input capacitor and device need to be sized to exhibit a large impedance at the frequencies of interest.

The input device is biased in the subthreshold region. Subthreshold conduction offers a number of benefits including a higher transconductance (i.e. more sensitive V-I transfer) and very low power consumption due to the much lower current (at the expense of slower response). The saturated drain current in the subthreshold region can be expressed as [18]

$$I_n = I_{D0}\left(\frac{W}{L}\right)_n \exp\left(\frac{V_{GS}}{nV_T}\right) \tag{3.2}$$

Table 3.2 RF detectors in literature

	[8]	[9]	[11]	[12]	[14]	[16]	[17]
Technology	0.18 μm	0.18 μm	0.18 μm	0.18 μm	0.35 μm	65 nm	130 nm
Area (mm²)	–	0.0043	0.0016	0.06	0.03	–	0.045
Frequency (GHz)	1–5	$-(1)^a$	0.1–20	$-(5.2)^a$	0.9–2.4	$-(60)^a$	$-(70)^a$
Dynamic range	0.05–0.5 V_{amp}	0.05–0.6 V_{amp}	0.05–0.9 V_{amp}	0.05–1 V_{amp}	0.02–1 V_{amp}	0.03–0.4 V_{amp}	0.03–0.4 V_{amp}
Conversion gain	–0.85 V/V	–0.75 V/V	+0.7 V/V	23 mV/dBm	–50 mV/dBm	+0.72 V/V	+1 V/V
Loading	7.6 fF	21 fF	12 fF	–	13 fF	–	–
Power (W)	3.6 $μ^b$	0.6 m	10 $μ^b$	3.5 m	8.6 m	–	0.5 m

[a] A range is not specified
[b] Only detector core, without biasing circuits

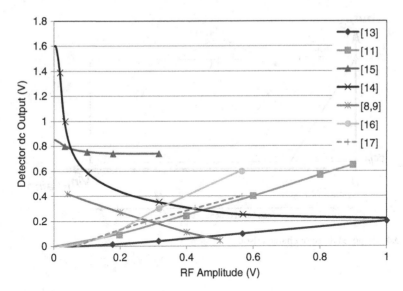

Fig. 3.2 RF-to-dc curves for various detectors in literature

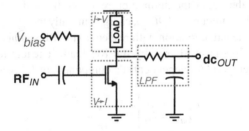

Fig. 3.3 Proposed detector architecture

where I_{D0} is a current constant independent of gate-to-source voltage, $(W/L)_n$ the aspect ratio of the NMOS transistor, n a process dependent term related to depletion region characteristics, and V_T the thermal voltage ($= kT/q$; 26 mV at room temperature).

With a small sinusoidal input $V_a cos(\omega t)$ superimposed on the gate bias V_{bias}, the following power series approximation of the current equation holds

$$I_n = I_{D0}\left(\frac{W}{L}\right)_n \exp\left(\frac{V_{bias}+V_a\cos(\omega t)}{nV_T}\right) = I_{B0}\exp\left(\frac{V_a\cos(\omega t)}{nV_T}\right)$$

$$\approx I_{B0}\left[1+\frac{V_a}{nV_T}\cos(\omega t)+\frac{1}{2}\left(\frac{V_a}{nV_T}\right)^2\cos^2(\omega t)\right]$$

$$= I_{B0}\left[1+\left(\frac{V_a}{2nV_T}\right)^2+\frac{V_a}{nV_T}\cos(\omega t)+\left(\frac{V_a}{2nV_T}\right)^2\cos(2\omega t)\right] \quad (3.3)$$

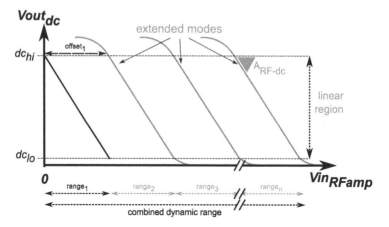

Fig. 3.4 Proposed detector characteristics

where $I_{B0} [=I_{D0}(W/L)_n exp(V_{bias}/nV_T)]$ is the dc-bias current of the transistor. The larger the amplitude, V_a, the larger the drained current and the further V_d is pulled away from V_{DD} causing the output node (DC_{out}), charged initially to V_{DD} by I_{bias}, to discharge thus establishing a negative relation with respect to the RF signal amplitude. As the detector output is low-pass filtered by the RC load, it reacts to the frequency-independent (but temperature dependent) dc component of I_n given by

$$I_{nDC} = I_{B0}\left[1+\left(\frac{V_a}{2nV_T}\right)^2\right]$$ (3.4)

This discharging dc current then becomes a dc voltage at the load. Proper sizing of the load and the input device will enable a high RF-to-dc conversion gain; however, with a limited voltage supply, increasing the conversion gain limits the dynamic range. To counter that, the detector is operated beyond its regular mode. By simply changing the gate bias of the input device while keeping the latter in the subthreshold region, the same detector characteristic can be extended to other amplitude ranges. This will essentially create extended operation modes with sub-ranged detection regions (Fig. 3.4). For higher amplitude signals, V_{bias} is made smaller such that the NMOS transistors are turned on with larger RF signal amplitudes. A digitally-programmable voltage bias circuit can then be implemented to shift the detector characteristic to the amplitude range of interest. The voltage shifts, or modes, can be designed to create continuous or overlapping regions with specified offsets. Upper and lower limits on the output dc value can then be used to automatically shift to the previous or next mode, respectively. These limits can be chosen around the range where the conversion gains of all modes are identical. The combined response is capable of covering a wide range of amplitudes with a high conversion gain.

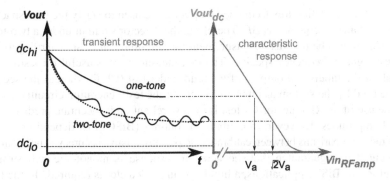

Fig. 3.5 One- and two-tone transient responses with all tones having same amplitude

Knowing the state of the detector, a dc output can be mapped to its corresponding RF amplitude reading as

$$V_{meas} = \frac{(dc_{hi} - dc_{out})}{|A_{RF-dc}|} + offset_x \tag{3.5}$$

where V_{meas} is the extracted reading, dc_{hi} the set limit of the detector output at zero signal, dc_{out} the detector's actual dc output, A_{RF-dc} the conversion gain, and $offset_x$ the offset between modes.

Of importance also is the applicability of two-tone signals in the BiST routines. Therefore, the behavior of the detector should be described under such inputs. Considering two closely spaced sinusoidal inputs $V_a cos(\omega_1 t)$ and $V_b cos(\omega_2 t)$ superimposed on the gate, several intermodulation products arise. However, due to the low pass nature of the load most of the non-linear components are attenuated except for those at the low delta frequency of $\omega_2 - \omega_1 (=\Delta F)$. In such case, the low frequency component of the discharging current I_n appearing at the output is then

$$I_{n\_low} = I_{B0}\left[1 + \left(\frac{V_a}{2nV_T}\right)^2 + \left(\frac{V_b}{2nV_T}\right)^2 + \left(\frac{V_a V_b}{(2nV_T)^2}\right)^2 + \frac{V_a V_b}{2(nV_T)^2}cos(\Delta Ft)\right] \tag{3.6}$$

The output is then a low frequency oscillating signal that can be easily digitized by the ADC. The average dc value of the oscillating output represents the contribution of two tones. If V_a and V_b are equal, the average dc output of the two-tone is approximately equivalent to the dc output of a single $\sqrt{2}V_a$ tone. That is the case for ideal two-tones, as is shown in Fig. 3.5. In a non-ideal case, the average dc output of the detector is also affected by the presence of other tonal elements, especially intermodulations of the third order that also appear close to the two tones. This forms the basic premise behind the detector's use for intermodulation distortion and linearity parameter extraction, to be discussed in Chap. 4.

Also, the oscillating low frequency output can be used to verify the tone spacing as its oscillation frequency is ΔF. Therefore, the detector's output under a two-tone test scenario can be used as a measurement for amplitude and also a verification of the test-signal's frequency spacing. The first measurement is useful for testing the signals at the inputs and outputs of a circuit-under-test (CUT) whereas the second can be used by the system as integrity check for the test-generation circuitry.

The use of an RF detector for test is mainly reliant on its accurate prediction of signal amplitudes. However, in calibration routines (BiSC) that depend on signal amplitude comparisons between calibration steps, only relative accuracy is necessary – meaning that a guarantee that the detector response is monotonic will suffice. Therefore for BiST applications, a highly accurate detector is required. To the first degree, it can be seen that the discharging current is a function of temperature and as such will create slight shifts in the characteristic response. Second, in smaller CMOS processes, variations in the devices severely affect the operation of the detector, for example the threshold voltage variations (see Table 1.2 in Chap. 1). Then, the detector would require calibration to characterize and stabilize its RF-to-dc conversion. For a fully embedded and standalone BiST approach, that calibration should be performed on-chip and not through separate testing.

In the next section, several implementations of the detector are described in various process nodes and for different high-frequency ranges. These implementations will showcase differential and single-ended designs aimed at both the microwave (RF) and millimeter wave bands. Also, a design that overcomes variations by self-adjusting and aligning its response is presented.

3.3.4 Implementations for RF and mm-Wave BiST

In this section we present a number of implementations of the proposed detector, namely a detector covering the 0.5–9 GHz range, 10–30 GHz range, and the 60 GHz band.

3.3.4.1 Microwave Implementation in 180 nm CMOS

Although an old technology by today's standards, 180 nm CMOS is still used for a number of RF applications as it is relatively inexpensive and able to cover a number of widely used standards such as Bluetooth, WiFi, DECT, and even any upcoming applications in TV whitespaces (e.g. IEEE 802.22 Wireless Regional Area Network).

The detector circuit is built in 180 nm CMOS technology from TSMC [19]. Figure 3.6 shows the detector's pseudo-differential core with an active pMOS load and output low pass filter. The input nMOS devices are chosen to be RF transistors and the input capacitors as MIM (Metal-Insulator-Metal).

The input stage devices are biased through large resistors by a separate programmable biasing circuit. The biasing circuit, shown in Fig. 3.7, is a simple two-bit voltage divider providing four discrete biasing voltages. The entire detector (core+biasing)

Fig. 3.6 Pseudo-differential
RF amplitude detector core

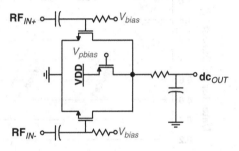

Fig. 3.7 Two-bit
programmable voltage
bias source

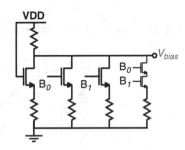

circuit runs from a 1.8 V supply and consumes less than 400 μA (~0.8 mW) when
in operation. To save on power, the detector's pMOS load can be turned off effectively
shutting down the core.

The characteristic response of the detector under a differential 2.5 GHz RF input
signal is shown in Fig. 3.8. The zero-RF dc output was set at 1.6 V by design. It can
be seen that the four modes of operation cover a continuous range from 0 to 0.7 V
amplitudes with each mode having a linear region between 1.6 V on the upper end
and 0.2 V on the lower. Therefore, these two endpoints are used to construct the
combined response as shown in Fig. 3.9, revealing a conversion gain of −10 V/V.
The detector also exhibits a very broadband range of frequencies, extending from
500 MHz to 9 GHz, over which the response holds constant. Moreover, the detector's
input impedance is larger than 8.5 kΩ over the frequencies of interest.

Of note here is the deviation of the dc response at very low signal amplitudes.
It can be observed that the linear relationship does not hold at signal amplitudes
below 100 mV where it becomes a square relation, as depicted in Fig. 3.10. This
discrepancy should be taken into account when processing the detector's output.
For example, when operating in the first mode and the detector's dc output is greater
than 1.35 V, then the linear approximation of (3.5) does not completely hold and
needs minor modification. The curve can be split into two at the dc output corresponding to 100 mV RF input, dc_{sq}, and the following can be used to determine the
measured RF amplitude,

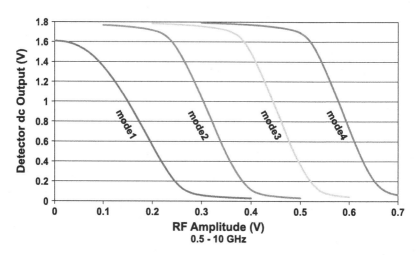

Fig. 3.8 Response of microwave amplitude detector at 2.5 GHz

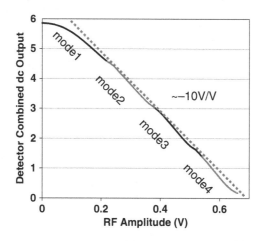

Fig. 3.9 Combined continuous linear response of the detector for input frequencies between 0.5 and 9 GHz

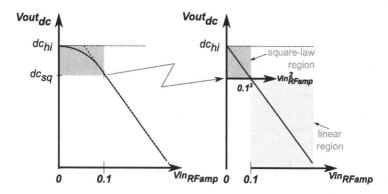

Fig. 3.10 Square-law and linear regions of the detector response at low RF signal amplitudes

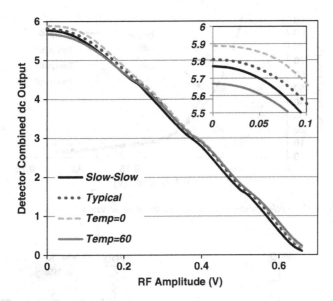

Fig. 3.11 Deviations of the detector's response due to temperature and process variations

$$
V_{meas}\big|_{offset_0} = \begin{cases} \sqrt{\dfrac{(dc_{hi} - dc_{out})}{\left|100 \left(dc_{hi} - dc_{sq}\right)\right|}} & when\ dc_{out} \geq dc_{sq} \\[4mm] \dfrac{(dc_{sq} - dc_{out})}{\left|A_{RF-dc}\right|} + 0.1 & when\ dc_{out} < dc_{sq} \end{cases} \tag{3.7}
$$

The frequency-independent response however is not immune to process and temperature changes. Figure 3.11 shows the response curves for a number of possible variations in operating conditions. The deviation from the ideal case, which is used to build the baseline dc-to-amplitude mapping function, will then result in very large amplitude measurement errors. One simple and passive method to partly remedy this issue is to perform a one-point calibration by re-referencing the zero-RF dc output to the ideal value (dc_{hi}), i.e. "zero-RF re-referencing". For example, the zero-RF dc output at 60°C is 1.47 V rather than the ideal 1.6 V. Knowing that, the 1.47 V can be virtually referenced as 1.6 V in the DSP after the RF detector's output is digitized by the ADC. Figure 3.12 shows the measurement accuracy with and without zero-RF re-referencing for the first mode of the detector. It can be seen that measurement accuracy is increased by limiting the dc-to-amplitude mapping error to less than 10% at low amplitudes and less than 5% onwards. While this technique works on this CMOS node, it might not be as easy on processes that exhibit much larger variations. Therefore, a method that can stabilize the detector's response with PVT is presented in Sect. 3.3.4.3.

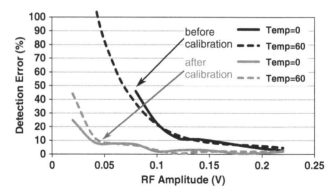

Fig. 3.12 Zero-RF re-referencing to increase measurement accuracy

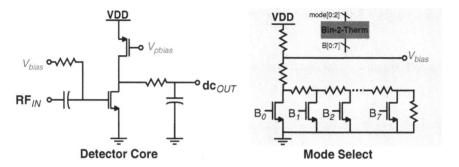

Fig. 3.13 mm-Wave amplitude detector core and bias circuits

3.3.4.2   Millimeter-Wave Implementation in 90 nm CMOS

A single-ended implementation of the detector is also designed targeting the mm-wave spectrum (60 GHz ISM band) using 90 nm CMOS from IBM and powered by a 1.2 V supply. This implementation has a mode-select programmable bias achieving 8 overlapping modes. The detector core and biasing are shown in Fig. 3.13. The mode offsets are set to 50 mV. With a −9 V/V conversion gain, the covered amplitude range is between 0 and 0.5 V, as shown in Fig. 3.14. The upper and lower dc limits are set at 1 and 0.2 V, respectively. Similar to the earlier implementation, the mm-wave detector response is to a large extent frequency-independent showing the same characteristic for the 55–65 GHz range.

At extreme high frequencies such as the band under consideration, the capacitive load presented by the detector starts losing its high impedance. In lieu of input impedances in excess of 8–10 kΩ in the RF implementation, this mm-wave detector exhibits impedances at an order of magnitude less, between 800 and 1000 Ω. This still presents a small loading but should be accounted for in the design of the mm-wave system.

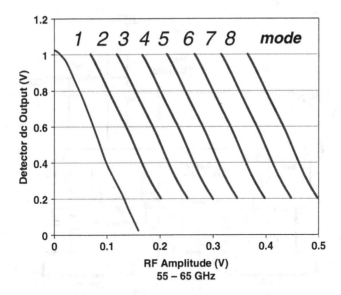

Fig. 3.14 mm-Wave sub-ranged amplitude detector characteristics

3.3.4.3   Self-Adjusting Detector Implementation in 65 nm CMOS

The one-point passive calibration method shown earlier to counter process and temperature mismatch does not track well with smaller CMOS processes where slight shifts in operating conditions throw the characteristic curve way off. A more involved calibration method is thus required.

There are two pivot points that can be used to stabilize the detector's characteristic curve. One is the zero-RF dc value, dc_{hi}, and the other is at the other end of the curve, i.e. at the last mode's dc_{lo}. If these two points can be fixed then the characteristic curve will be held in place with PVT variations.

Figure 3.15 shows the detector implementation, in 65 nm CMOS from IBM, for the 10–30 GHz frequency range with in-built self-adjustment capabilities. The design makes use of two replica cores, *zero-RF* and *max-RF* replicas, placed in proximity to the main core to reduce variations. Also, all the cores' input devices are sized at much larger than minimum length to have better matching between all copies, while keeping their input impedances relatively high (>1 kΩ at 30 GHz).

The first replica replaces the fixed programmable biasing of the previous implementations and works on fixing the upper dc output limit, dc_{hi}, to 1 V. Therefore, no RF signal is applied to that replica. Its output is compared to a fixed 1 V reference by an operational amplifier (op-amp), which through feedback adjusts that replica's gate voltage to force the appropriate bias point for dc_{hi}. An analog subtractor, described in Ref. [20], is inserted in the feedback path to enable a programmable mode-select. The latter takes the op-amp output and generates two bias voltages that are at a programmed offset from each other. This enables the main core to have a lower bias voltage for its extended modes while the replica core sets the relative starting bias voltage.

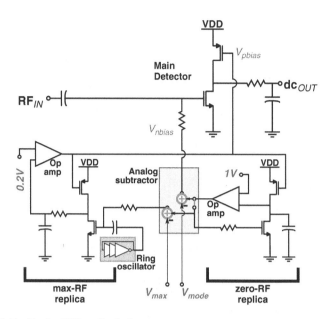

Fig. 3.15 Self-adjusting RF amplitude detector

The second point in need of stabilization is at the end of the characteristic curve. For that, another replica is included and is supplied the maximum RF amplitude – in this case, a rail-to-rail high-frequency oscillation. The max-RF replica is set at the last desired mode offset (V_{max}) also through an identical subtractor and its input is derived from a small inverter-based ring oscillator. The oscillator is built with a minimum sized three-inverter chain and buffered to the input of the max-RF core. The ring's frequency of oscillation is slightly higher than 15 GHz; but since the detector is broadband and frequency-insensitive, deviations in that frequency do not distract from the replica's functionality. The function of this replica is to adjust the loads of all cores such that its dc output at maximum RF input and last mode of operation is exactly dc_{lo}. This is forced through an op-amp regulating the pMOS load and comparing the max-RF replica's dc output to a 0.2 V reference.

The result is shown in Fig. 3.16 with changes in temperature, process, and input frequency resulting in extremely minimal and indiscernible shifts in the primary and extended modes of operation. The bias offset (V_{mode}) was selected to provide 4 slightly overlapping modes – whereby the last offset (the one applied to the max-RF replica, V_{max}) determines how steep the conversion gain is. Monte Carlo statistical simulations are also performed to enable a more realistic characterization of this method since mismatches between the cores, main and replicas, will affect the

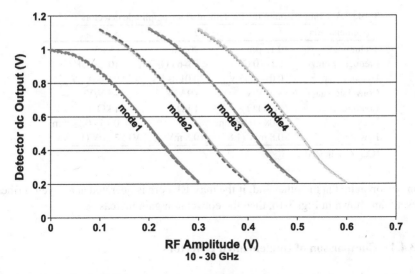

Fig. 3.16 Self-adjusting RF detector modes under different temperature and process variations

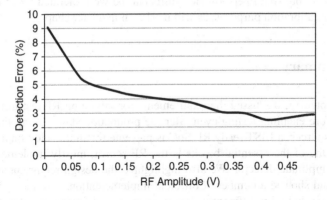

Fig. 3.17 Maximum detection errors (Monte Carlo simulations) across the amplitude range

detector's actual curves. More than 500 Monte Carlo runs with process and mismatch variations over a temperature range of −30 to 90°C show a maximum detection error of 8% at the lowest end, quickly reducing to less than 2% at the extended modes (Fig. 3.17).

One feature of this configuration is the ability to adjust the conversion gain. If, for example, both the main and max-RF cores are operated in the primary mode, then a single mode covers the 0–0.6 V RF amplitudes thereby decreasing the

Table 3.3 Comparison of the implemented RF amplitude detectors

Implementation	1	2	3
Technology	0.18 μm	90 nm	65 nm
Frequency range	0.5–10 GHz	55–65 GHz	10–30 GHz
Dynamic range	0.01–0.65 V_{amp}	0.01–0.5 V_{amp}	0.01–0.6 V_{amp}
Conversion gain	−10 V/V	−9 V/V	−5 V/V[a]
Loading	>8.5 kΩ	1 kΩ	1 kΩ
Area	0.08×0.08 mm²	–	0.15×0.10 mm²
Power	0.8 mW (1.8 V)	0.5 mW (1.2 V)	3 mW (1.2 V)

[a]Can be adjusted

conversion gain. On the other end, if the max-RF core is operated at an even further offset than shown in Fig. 3.16, then the conversion gain increases.

3.3.4.4   Comparison of Implemented Detectors

The specifications of the designed RF and mm-wave amplitude detectors are summarized in Table 3.3. These detectors will be used in the later chapters for BiST and BiSC. As stated previously, a highly accurate implementation is preferred in the case of testing but small performance shifts can be well tolerated when using the detector for calibration purposes, as will be shown in a later chapter.

3.4   Summary

In this chapter, we discussed the requirements for migrating RF test to inside the chip therefore enabling testing even after the production stage. Using the various on-chip resources, a BiST-ready RF SoC is presented with the additional enabling circuitry. One of the essential blocks is the RF sensor, mostly implemented as a power or amplitude detector. We also highlighted the design requirements for RF detectors and showed a number of different implementations. We then proposed a similar design but with different capabilities, enhanced for true on-chip testing in RF SoC. Three implementations in different CMOS technologies and covering the RF and mm-wave bands are discussed and their performances compared. In the next chapter, we will put this detector to use in various testing schemes and routines.

References

1. F. F. Dai, C. Stroud, D. Yang, "Automatic linearity and frequency response tests with built-in pattern generator and analyzer," *IEEE Transactions on Very Large Scale Systems (VLSI)*, vol.14, no.6, pp.561–572, June 2006
2. M. Onabajo, J. Silva-Martinez, F. Fernandez, E. Sanchez-Sinencio, "An On-Chip Loopback Block for RF Transceiver Built-In Test," *IEEE Transactions on Circuits and Systems II: Express Briefs*, vol.56, no.6, pp.444–448, June 2009

3. R. Ramzan, J. Dąbrowski, "CMOS blocks for on-chip RF test," *Analog Integrated Circuits and Signal Processing*, vol.49, no.2, pp.151–160, 2006

4. J.-S. Yoon, W. R. Eisenstadt, "Embedded loopback test for RF ICs," *IEEE Transactions on Instrumentation and Measurement*, vol.54, no.4, pp.1715–1720, Oct. 2005

5. R. Ramzan, S. Andersson, J. Dabrowski, C. Svensson, "Multiband RF-Sampling Receiver Front-End with On-Chip Testability in 0.13 μm CMOS," *Analog Integrated Circuits and Signal Processing*, vol.61, no.2, pp.115–127, Feb. 2009

6. A. Valdes-Garcia, J. Silva-Martinez, E. Sanchez-Sinencio, "On-Chip Testing Techniques for RF Wireless Transceivers," *Design & Test of Computers, IEEE*, vol.23, no.4, pp.268–277, April 2006

7. J.J. Dabrowski, R. Ramzan, "Built-in Loopback Test for IC RF Transceivers," *IEEE Transactions on Very Large Scale Integration (VLSI) Systems*, vol.18, no.6, pp.933–946, June 2010

8. F. Jonsson, H. Olsson, "RF detector for on-chip amplitude measurements," *Electronics Letters*, vol.40, no.20, pp.1239–1240, Sept. 2004

9. C. Zhang, R. Gharpurey, J.A. Abraham, "Low Cost RF Receiver Parameter Measurement with On-Chip Amplitude Detectors," *VLSI Test Symposium, IEEE, 26th IEEE VLSI Test Symposium*, pp. 203–208, 2008

10. A. Valdes-Garcia, R. Venkatasubramanian, R. Srinivasan, J. Silva-Martinez, E. Sanchez-Sinencio, "A CMOS RF RMS Detector for Built-in Testing of Wireless Transceivers," *VLSI Test Symposium, IEEE, 23 rd IEEE VLSI Test Symposium (VTS'05)*, pp.249–254, 2005

11. Q. Wang, M. Soma, "RF Front-end System Gain and Linearity Built-in Test," *VLSI Test Symposium, IEEE, 24th IEEE VLSI Test Symposium*, pp.228–233, 2006

12. H.-H. Hsieh, L.-H. Lu, "Integrated CMOS Power Sensors for RF BIST Applications," *VLSI Test Symposium, IEEE, 24th IEEE VLSI Test Symposium*, pp.234–239, 2006

13. Y.-C. Huang, H.-H. Hsieh, L.-H. Lu, "A Built-in Self-Test Technique for RF Low-Noise Amplifiers," *IEEE Transactions on Microwave Theory and Techniques*, vol.56, no.5, pp.1035–1042, May 2008

14. A. Valdes-Garcia, R. Venkatasubramanian, J. Silva-Martinez, E. Sanchez-Sinencio, "A Broadband CMOS Amplitude Detector for On-Chip RF Measurements," *IEEE Transactions on Instrumentation and Measurement*, vol.57, no.7, pp.1470–1477, July 2008

15. F. Xiaohua, M. Onabajo, F.O. Fernandez-Rodriguez, J. Silva-Martinez, E. Sanchez-Sinencio, "A Current Injection Built-In Test Technique for RF Low-Noise Amplifiers," *IEEE Transactions on Circuits and Systems I: Regular Papers*, vol.55, no.7, pp.1794–1804, Aug. 2008

16. J. Gorisse, A. Cathelin, A. Kaiser, E. Kerherve, "A 60 GHz CMOS RMS power detector for antenna impedance mismatch detection," *NEWCAS-TAISA 2008. 2008 Joint 6th International IEEE Northeast Workshop on Circuits and Systems and TAISA Conference, 2008*, pp.93–96, 22–25 June 2008

17. S. Rami, W. Tuni, W. R. Eisenstadt, "Millimeter wave MOSFET amplitude detector," *2010 Topical Meeting on Silicon Monolithic Integrated Circuits in RF Systems (SiRF)*, pp.84–87, Jan. 2010

18. D.J. Comer, D.T. Comer, "Using the weak inversion region to optimize input stage design of CMOS op amps," *IEEE Transactions on Circuits and Systems II: Express Briefs*, vol.51, no.1, pp.8–14, Jan 2004

19. Taiwan Semiconductor Manufacturing Company, Hsinchu, Taiwan, www.tsmc.com

20. R. Fried, C.C. Enz, "Simple and accurate voltage adder/subtractor," *Electronics Letters*, vol.33, no.11, pp.944–945, May 1997

Chapter 4
RF Built-in-Self-Test

This chapter expands on the previous discussion on the BiST-ready RF SoC and the design of suitable RF sensors for true on-chip test. We place the RF amplitude detector designed in the previous chapter in the test loop to enable a number of important characteristic tests for RF and mm-wave blocks. We first describe the overall BiST routine that makes use of the digital core, transmitter circuits and loopback elements to generate test signals and route them accordingly to accomplish specification testing of most RF blocks through simple one- and two-tone tests. Then we demonstrate the effectiveness and viability of the RF detector in predicting and quantifying signal and circuit parameters through simulation examples in various frequency bands and process technologies.

4.1 Specification-based Tests Using the RF Amplitude Detector

In this section, specification-based testing of various RF blocks is described with the use of the developed RF detectors. Signal amplitude can be indicative of various RF block parameters and as such its proper detection and extraction by means of an accurate detector enables the implementation of on-chip self-test. It was shown in the previous chapter that the detectors have an inverse relationship to the RF amplitude, i.e. the dc output decreases with increased RF amplitude. The detector responses are used to map a dc value to a corresponding RF amplitude, in other words amplitude prediction or extraction.

Here we describe the methods by which the detector can be used to extract RF block performance parameters such as gain, compression point, intermodulation distortion, and quadrature mismatch. The BiST-ready SoC architecture presented in Chap. 3 is repeated here (Fig. 4.1) with node numbers for convenience and ease of referencing.

Depending on the block being measured – the circuit under test (CUT) – some bypasses and attenuation might be required. The detectors can be also used to ensure that the appropriate test signal is being routed to the CUT. Care should be taken to provide the CUT with suitable test signals, for example ones that do not send it into

S. Bou-Sleiman and M. Ismail, *Built-in-Self-Test and Digital Self-Calibration*
for RF SoCs, SpringerBriefs in Electrical and Computer Engineering,
DOI 10.1007/978-1-4419-9548-3_4, © Springer Science+Business Media, LLC 2012

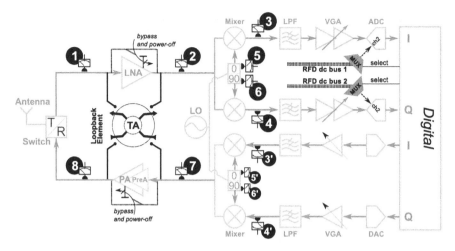

Fig. 4.1 BiST-ready SoC architecture with the important measurement points along the transceiver

Table 4.1 Test setups for the various RF blocks

Circuit under test	Loopback connect	Disable	Nodes to monitor
LNA	$(7) \Rightarrow (1)$	PA	(1) and (2)
	$(8) \Rightarrow (1)$	–	
PA	No loopback	–	(7) and (8)
Downconversion mixer	$(7) \Rightarrow (2)$	LNA	(2)–(6)
	$(8) \Rightarrow (2)$	LNA	
LO	No loopback	–	(5) and (6)
Upconversion mixer	No loopback	–	(3′), (4′), (5′), (6′), (7)
Test attenuator and switches	$\binom{7}{8} \Rightarrow \binom{1}{2}$	–	(1), (2), (7), (8)

compression, for certain types of tests. Therefore, sufficient attenuation can be achieved by controlling the test-generation and test-routing circuitry. For testing the PA, the previous blocks should have tunable components such as the upconversion mixer or LO – with preference on the earlier. For the receiver chain, for example, the test attenuator (and offset mixer, if any) should be tuned accordingly. The tuning of these test-generation and test-routing elements can be monitored and controlled because of the presence of amplitude detectors at their nodes – rendering them like any other CUT.

The BiST routine starts with creating the required test signal in baseband and upconverting to RF. The amplitude of the test signal can be monitored at node (7), or node (8), and adjusted if a tunable mixer is implemented. Testing the test-generation circuitry increases confidence in the overall setup and test results. For testing the receiver, the loopback element is activated and depending on the type of test and signal level, the test signal can be routed from (7) or (8) to nodes in the receiver. Table 4.1 lists the various circuits under test and their testing setups while stating which nodes to connect and to monitor, and what blocks can be turned off.

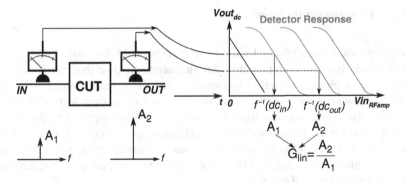

Fig. 4.2 Setup for one-tone test for gain measurement

Not only can individual circuits be tested but also a cascade test can be set up with the appropriate connections and nodes. Next we describe the various tests that can be run with the above setups such as gain, linearity, intermodulation distortion, and quadrature mismatch.

4.1.1 Gain

To extract the gain of an RF block, its input and output nodes have to be monitored. A one-tone test signal can be generated by the test-generation circuitry in the SoC and routed to any block along the loopback chain. Since the signals are measured at the boundary of the RF circuit, the block's parameters are properly extracted thus eliminating any forms of performance masking that might appear in cascaded chains.

In the previous chapter, the detector's accuracy (after some calibration) is shown to be within 5–10%, which translates to a gain prediction error of less than 1 dB using only on-chip components. Figure 4.2 depicts the gain measurement under a one-tone test. The predicted gain is then

$$G_{lin} = f^{-1}\left(dc_{out}\right)/f^{-1}\left(dc_{in}\right) \longrightarrow A_2 / A_1$$
$$G_{dB} = 20\log\left(G_{lin}\right)$$

(4.1)

where $f^{-1}(dc)$ is the mapping function relating the detector's dc output to RF amplitude, dc_{in} the output of the input detector, dc_{out} the output of the output detector, A_1 the amplitude of the input one-tone, and A_2 the amplitude of the output tone.

Gain measurements for the mixer are a bit different as its inputs and outputs are at different frequencies. The detector can be leveraged to measure the amplitudes at the high frequency ports only, while a baseband sensor should measure the low frequency ports. Conversion gain can then be deduced from both sensors' readings.

4.1.2   Compression Point

The compression point (P_{1dB}) of a CUT can be extracted following a one-tone test sweep. Using the same setup as the gain measurements described previously, the input signal amplitude is changed in discrete steps and the gain recorded. The input amplitude that results in a reduction of the gain by 1 dB is then deemed the compression point. The discrete amplitude levels can be either supplied by the upconversion mixer for the transmitter tests, in addition to the loopback elements for the receiver tests. The gain points gathered are used to construct the gain curve, as a piecewise fitting curve, and find the input amplitude (and power) that results in a 1 dB reduction in gain. Therefore,

$$G_{lin}[A_{1,x}] = f^{-1}\left(dc_{out,x}\right)\Big/f^{-1}\left(dc_{in,x}\right)$$

$$P_{1db} = \left(A_{1,y}\right)_{dBm}\Big|_{G_{dB}[A_{1,0}]-G_{dB}[A_{1,y}]=1} \tag{4.2}$$

where x is the iteration step and A_1 the amplitude-swept input tone; $A_{1,y}$ represents the amplitude that results in compressing the gain, with y not necessarily being one of the discrete set of points {x}, i.e. it can be extrapolated between two discrete points.

4.1.3   Intermodulation Distortion

To test for intermodulation distortion, or more specifically IIP3, a two-tone test is needed. Upon its generation by the transmitter baseband and upconversion, the two-tone signal can be applied at various nodes in the system. Loopback elements have very good linearity metrics of their own, by design, to keep the test signal from getting affected.

In Chap. 3, the response of the detector under a two-tone test is shown. While this is not a pure dc signal, the basic premise here is that averaging that low frequency signal is a task easily accomplished in the DSP following the ADC. The average corresponds to a composite amplitude of the tones present in the signal. The two-tone signal without any distortion should result a detector dc output mapped to exactly 1.414 times a single tone's amplitude (A_1). If a clean two-tone is assumed at the input, and the CUT gain (G) has been tested for and known, then the expected output amplitude should map to exactly a linear scaling of the input tones ($\sqrt{2}A_2 = \sqrt{2}GA_1$). The emergence of third order intermodulation distortion results in a detected amplitude in excess of the expected scaling. The difference between the detected and expected output amplitude is traced to the contribution of the new tones in the spectrum, the IM3 amplitudes (B), as depicted in Fig. 4.3. That is, given the gain of the CUT (G), the following holds

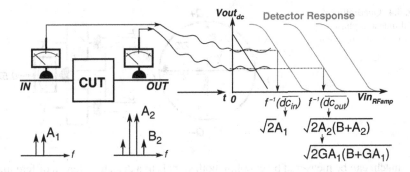

Fig. 4.3 Two-tone test setup and IM3 measurement

$$f^{-1}\left(\overline{dc_{in}}\right)=\sqrt{2A_1^2}$$

$$f^{-1}\left(\overline{dc_{out}}\right)\approx\sqrt{2A_2(A_2+B)}$$

$$B\approx\frac{\sqrt{2}}{2}G\times f^{-1}\left(\overline{dc_{in}}\right)\times\left(\left(\frac{f^{-1}\left(\overline{dc_{out}}\right)}{G\times f^{-1}\left(\overline{dc_{in}}\right)}\right)^2-1\right)$$

$$(4.3)$$

where the averages of dc_{in} and dc_{out} are mapped through $f^{-1}(dc)$ and A_1 and A_2 are the two-tone signals at the input and output respectively, and B the intermodulation distortion amplitude. The two-tone test signal power at the input needs to be below the compression point of the CUT for the above to provide accurate parameter extractions.

Again, the mixer is a special case. For example, a downconversion mixer's IIP3 and IIP2 can be measured with the RF detector and a baseband detector. Two-tone test signals are also required from the test generation circuitry. The two-tone signals need to be carefully crafted to place either a second or third order intermodulation distortion product inside the output low-pass filter's bandwidth. That component can then be detected by the baseband and used to quantify the IIP3 or IIP2 of the mixer in conjunction with the RF input amplitude, sensed by the RF detector, as in Ref. [1].

4.1.4 Quadrature Mismatch

Amplitude and phase mismatches in the quadrature modulation and demodulation results in incorrect reception and transmission of symbols. A primary cause of these mismatches is the local oscillator (LO). The high-frequency LO quadrature outputs, I and Q, are ideally of equal amplitudes and exactly 90° apart. The RF detector can be used here to obtain these mismatches by monitoring these two outputs.

Detecting amplitude mismatch is trivial as it can be directly spotted if there is a difference in the outputs of the detectors monitoring the I and Q signals. It can then be easily quantified. The phase between the quadrature signals, and hence the phase

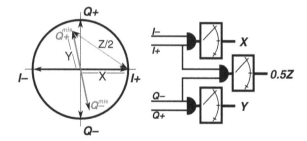

Fig. 4.4 Quadrature amplitude and phase mismatch measurement

mismatch, can be measured by coupling both signals to a detector. Here, a differential detector is needed as well as differential quadrature LO, which is the most common implementation with the use of fully balanced differential mixers. By connecting one branch of the I signal and another from the Q signal to the differential input stage of the detector, a hybrid equivalent signal is sensed. A polar diagram that best demonstrates this detection method is shown in Fig. 4.4. When sensing each path by itself, the respective detectors' dc outputs correspond to the amplitudes of vectors X and Y. If the detectors' outputs are different, then X and Y are amplitude mismatched. However, when sensing the hybrid signal, the detector's dc output corresponds to the resultant vector Z. Knowing the amplitudes of X, Y, and Z, the cosine law enables the extraction of the phase between the I and Q paths. Then phase mismatch is the deviation of that angle from 90° as described in

$$\overbrace{f^{-1}\left(dc_I\right)=X \quad f^{-1}\left(dc_Q\right)=Y}^{amplitude\ mismatch=\alpha=X/Y} \quad \underbrace{f^{-1}\left(dc_{I/Q}\right)=Z/2}_{phase\ mismatch=\varphi=90-\frac{180}{\pi}\cos^{-1}(\frac{X^2+Y^2-Z^2}{2XY})} \tag{4.4}$$

4.1.5 Isolation and Feedthrough

Some elements in the transceiver should exhibit very good isolation between their ports. For example, a MOS switch in the off-state should ideally present infinite isolation. However, parasitics can couple part of the signal at one end of the switch to the other end. This is quite important in the loopback scheme discussed here, where the loopback element contains routing switches that on one end, at the transmitter interface, have high power signals and on the other end at the receiver interface, weak and low power signals. The RF detector can then be used to detect these RF components across an off-state switch. In such case, the one-tone test setup as depicted in Fig. 4.2 is used to quantify the extent of isolation, or lack thereof. A high-powered input test signal is preferred as it enables better visibility and detectability of an attenuated signal at the output.

Mixers also should exhibit a level of isolation to prevent the high-power LO signal from appearing at the IF or RF ports. This LO feed through is detrimental in both downconversion and upconversion mixers. In downconversion mixers, self-mixing

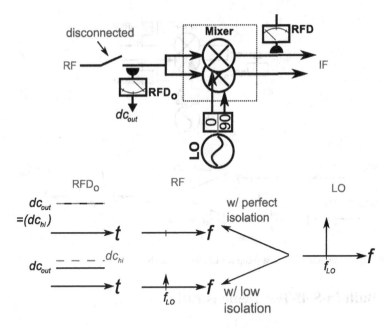

Fig. 4.5 Downconversion mixer test setup for LO to RF port isolation

due to the coupling of the LO signal to the input RF port results in an undesirable dc offset. To test for this, the setup shown in Fig. 4.5 can be implemented with the input RF port silenced (by disconnecting it from the other circuits, e.g. turning off the LNA and loopback) while the LO still drives its own port. If coupling is extremely low, then the RF amplitude detector will stay at its zero-RF dc value (dc_{hi}). However, if the LO couples to the RF port, then it will excite the RF detector at that port that in turn will detect a high-frequency amplitude. Since no other RF input is present, then this detection quantifies the amount of LO coupling and subsequent self-mixing.

In the transmitter, the upconversion mixer's LO feed-through (from the LO port to the output RF port) is problematic as it can fall in the vicinity or directly on top of the desired tone. One method to detect LO feed-through at the RF port of the upconversion mixer is to supply a low frequency tone at the transmit IQ baseband, and monitor the RF output with the RF amplitude detector. Without LO feed through, a single tone should be present at the output and the detector's response will settle to a dc value corresponding to that signal's amplitude. If an LO tone is also present, then the detector's behavior resembles the two-tone case where its output is an oscillating signal at the frequency of the frequency offset between the LO and the upconverted test tone (see Equation (3.6) and Fig. 3.4). A visualization is shown in Fig. 4.6. For best LO visibility at the RF port, a smaller baseband test tone is preferred in order not to mask the coupled LO signal. Also it is assumed that no IQ mismatch is present in this setup as that will create also an image tone symmetric to the RF tone with respect to the LO frequency.

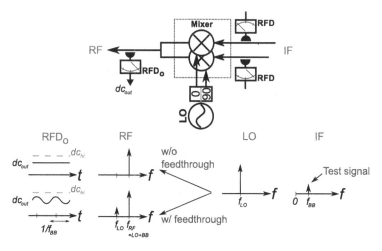

Fig. 4.6 Upconversion mixer test setup for LO feedthrough

4.2 Built-in-Self-Test Demonstration

In this section, we demonstrate the use of the detector in the previously described routines to characterize RF circuits and signal phenomena. Test-benches are first used to verify the usability of the detector in extracting some signal phenomena, irrespective of the circuitry it is attached to. This allows more controllability on the applied signals and provides a proof of the detector's viability for such tests and signals. Then, actual circuits-under-test are monitored and their extracted parameters are compared to their simulated ones. Two LNAs are built: one operating at 2.4 GHz in 180 nm CMOS and another operating at 60 GHz in 90 nm CMOS. The LNA is selected since it is one of the most critical circuits in the transceiver – and the RF detector is used to quantify its gain and linearity.

4.2.1 Detector Test-Benches

Before connecting the detector to the RF circuits, we verify here the effectiveness of the detector in quantifying some of the signal phenomena that appear in transceivers, namely intermodulation distortion and phase mismatch. The signals are applied directly to the RF detector and the output is digitized using a ten-bit ADC and mapped to an amplitude measurement. In the following test-benches, the RF implementation of the amplitude detector (180 nm) is used, as described in Chap. 3.

4.2.1.1 Intermodulation Distortion

To test the intermodulation distortion characterization method described in the earlier section, we implement a test-bench with the RF detector being supplied

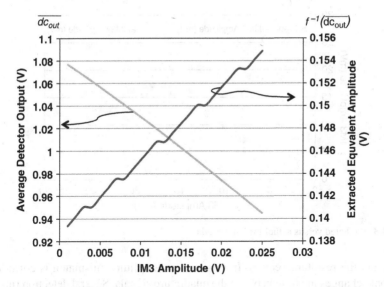

Fig. 4.7 RF detector output and mapped output amplitude in response to a two-tone input with varying IM3 component

with a controlled two-tone signal. The amplitude of a single tone is set at 0.1 V (A_I). Intermodulation components are added to the original signal and the output of the detector is observed and used to map to the corresponding amplitudes. As described earlier, the detector's output in a two-tone test is a low frequency oscillating signal of which the average is used to compute signal amplitudes.

It is expected then that a clean two-tone stimulus to the detector should be interpreted as $1.4A_I$, or in this case 0.14 V. Also, with increasing the injected IM3 component, the equivalent extracted amplitude (which is a function of the average dc output of the detector) should increase. The results are shown in Fig. 4.7, where the average dc output of the detector (at 1.08 V) corresponds to around 0.14 V (RF amplitude) when no distortion tones are present. Subsequent increases in the distortion tones (IM3) result in a decrease of the detector output and a resulting reciprocal behavior in the extracted amplitude.

Processing this information, we are able to extract the IM3 contribution given the offset created by the addition of that tone, in accordance with the previously established mappings in Equation (4.3). Figure 4.8 plots the predicted IM3 amplitudes versus the actual amplitudes (diagonal) showing very good correspondence.

4.2.1.2 Quadrature Phase Mismatch

In this test, we demonstrate the extraction of the phase mismatch between two differential quadrature signals – I and Q. Three differential RF amplitude detectors are used where the differential I signal, Q signal, and the composite single-ended I/single-ended Q signal are fed to the first, second, and third RF amplitude detectors, respectively. These detectors will measure the amplitude components related to the

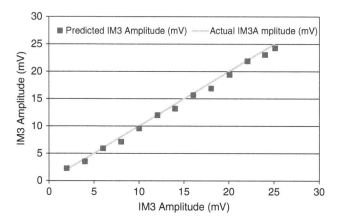

Fig. 4.8 Predicted versus actual IM3 amplitude

I, Q, and the resultant vectors. In this test, no amplitude mismatch is considered and only changes in phase between the quadrature signals. Several detection runs are performed with phase mismatches between $-10°$ and $10°$. Referring to Fig. 4.4, the RF detector connected to the differential I signals records an amplitude X, while the one connected to the differential Q signals records a Y amplitude. The composite amplitude Z is also predicted. The phase imbalance is calculated using these amplitude measurements as described in Equation (4.4).

Figure 4.9 shows the results of the test runs with the predicted phases aligning with the actual phase shifts. The prediction error is within $1°$ for a ten-bit quantized detector output.

4.2.2   LNA as Circuit-Under-Test

In this section, we use the detector to characterize two LNAs at two different frequency bands. The one-tone and two-tone tests described in the earlier section are used to provide gain and linearity measurements. Also it is shown that the detector does not load the circuit-under-test. Detectors are placed at the input and output of the LNAs and their modes are selected depending on their dc output limits.

4.2.2.1   2.4 GHz LNA

A 2.4 GHz LNA is built in 180 nm CMOS from TSMC and used with the RF amplitude detector. We will leave the implementation and circuit of the LNA for the next chapter. The input one-tone is varied and the inputs and outputs of the LNA are observed and detected. The dc outputs of the detectors are quantized using an eight-bit ADC and are used to obtain the gain. Figure 4.10 shows the results of the input sweep with the extracted gain matching to within 0.5 dB of the real gain.

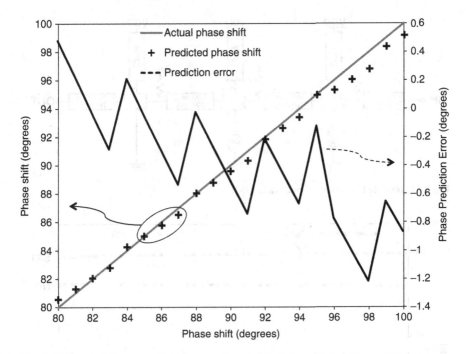

Fig. 4.9 Phase imbalance prediction versus the actual imbalance (*left axis*); prediction error in degrees (*right axis*)

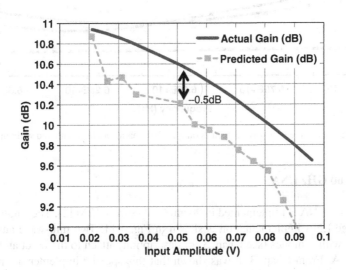

Fig. 4.10 2.4 GHz LNA gain extraction: actual versus predicted gain curve

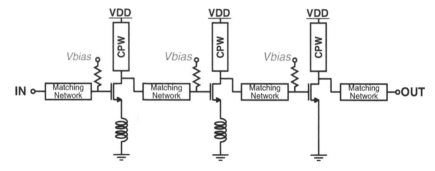

Fig. 4.11 60 GHz LNA used for mm-wave BiST

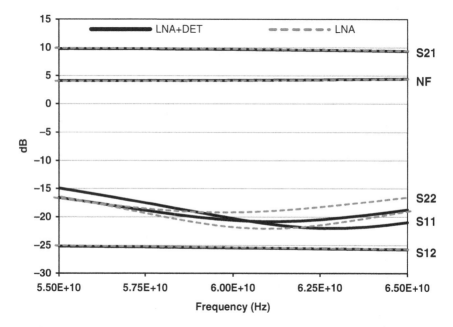

Fig. 4.12 Effect of the mm-wave detectors on the LNA characteristics: with and without

4.2.2.2 60 GHz LNA

A mm-wave LNA is implemented in 90 nm CMOS from IBM [2]. The single-ended three-stage LNA implementation is shown in Fig. 4.11. Two mm-wave amplitude detectors with 8-modes (see Chap. 3 for description) attach to the input and output of the LNA. From Chap. 3, it was shown that this specific implementation of the detector presents around 1 kΩ impedance and will therefore impact the LNA minimally. Figure 4.12 shows the LNA s-parameters with and without the connected mm-wave

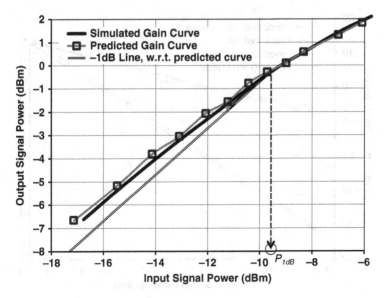

Fig. 4.13 Actual and predicted gain curve after one-tone test sweep

detectors. It can be seen that the detectors do not cause degradation in the LNA's original performance.

One-tone and two-tone test sweeps are performed on the LNA with the detectors. The outputs of the detectors are quantized by an eight-bit ADC and the digital word is mapped to an extracted amplitude measurement. After a number of one-tone iterations, the gain curve can be constructed and the P_{1dB} point deduced. This is shown in Fig. 4.13 with the gain compression point highlighted.

Also two-tone test sweep is performed with a clean two-tone signal at the input of the LNA. The input and output detectors' average dc output is again mapped to a resulting amplitude that can be used to extract the properties of the signals. Since the gain is known from the previous test, the intermodulation distortion BiST routine described earlier in the chapter is used to detect the emergence of third order intermodulations at the output of the LNA. Figure 4.14 shows the results with the extrapolated curves for the simulated and predicted values. Table 4.2 summarizes the actual and extracted parameters of the LNA with the on-chip BiST matching the gain to within 0.3 dB error, the 1 dB compression point and IIP3 to within 0.4 dB.

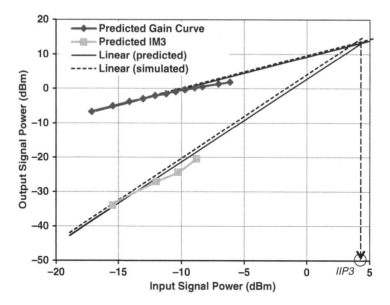

Fig. 4.14 Third order intermodulation extraction and IIP3 measurement

Table 4.2 Actual versus predicted parameters for the 60 GHz LNA

LNA parameter	Actual	Predicted	Error
Gain, dB	10.14	10.45	0.3
P_{1dB}, dB	−9.73	−9.32	0.4
IIP3, dB	3.9	4.3	0.4

4.3   Summary

In this chapter, we expanded on the topic of Built-in-Self-Test as it applies to RF SoCs and described specification-based tests for the extraction of RF block parameters. We also placed the RF amplitude detector into the testing routine and used its capabilities to measure a number of circuit parameters such as gain, compression, and non-linearity, as well as amplitude and phase mismatch in quadrature signals.

These tests enable system self-awareness taking the system one step forward towards self-healing and self-calibration. Next chapter will describe the use of the amplitude detector in calibration routines and showcase a few circuit examples.

References

1. S. Rodriguez, A. Rusu, L.-R. Zheng, M. Ismail, "A Novel BiST and Calibration Technique for CMOS Down-Converters," *ICCSC 2008. 4th IEEE International Conference on Circuits and Systems for Communications, 2008*, pp.828–832, May 2008
2. A. Akour, M. Ismail, R. Rojas-Terran, "Manufacturable 60 GHz CMOS LNAs," *Proceedings of the SDR '08 Technical Conference and Product Exposition*, October 2008

Chapter 5
RF Built-in-Self-Calibration

The notion of cognitive radios came about a bit more than a decade ago, where it was suggested the radios should become powerful enough and smart enough to sense the outside world and environment and adjust their inputs and outputs accordingly. However, this deals only with the outside world, introspect cognition is also needed – an awareness of the radio's own state – to allow it to adjust, or calibrate, its own internal components, especially now that the radio building blocks are becoming more volatile and prone to failure.

Calibration is usually done in the factory before an item is shipped and stored in memory, but this comes at an increased cost to the manufacturing and testing budgets. Self-calibration can be built into the system to be run at system start-up or at predetermined and periodic occasions, for example when a self-test deems that performance has noticeably degraded. The advantage of having a baseband DSP is a major enabler in making calibration self-contained and embedded within the system requiring no off-chip or off-system assistance.

For calibration to apply to RF blocks, they have to be designed and built with tunable elements in order to be programmed and tuned, on-the-fly. However, a key component here is self-awareness, on the block and system levels. A system can gain self-awareness if proper testing and monitoring procedures are also part of its capabilities. In the previous chapters, we showed how a RF SoC can be made self-test-ready with the addition of loopback elements and most importantly accurate RF sensors. In this chapter, we take these capabilities and enhance them with self-adjusting capabilities of RF blocks, in line with the digitally-assisted RF design principles.

First, we describe how and why digital calibration is the most viable solution to analog impairments and highlight some of the self-calibration design requirements. Also, we show examples of digitally-assisted RF circuits and their calibration routines.

S. Bou-Sleiman and M. Ismail, *Built-in-Self-Test and Digital Self-Calibration for RF SoCs*, SpringerBriefs in Electrical and Computer Engineering, DOI 10.1007/978-1-4419-9548-3_5, © Springer Science+Business Media, LLC 2012

5.1   RF Block and System Self-Awareness

RF designers deal with an interesting challenge when building circuits in nanoscale CMOS: the integrated devices they use are now capable of offering good high-frequency performance only to suffer from very low reliability. The margin in which an RF circuit achieves its best performance is not only very narrow but also quite variable. The operating points that yield the optimal performance are not fixed; they are a function of many variables, including process, voltage and temperature. The increased susceptibility of individual transistors, the basic circuit building blocks, to sway away from their set characteristics makes the successful implementation of even the smallest and simplest of RF circuits an exercise in probability. While overdesign reigned in variability in older technologies, it simply cannot overcome the new challenges. One method is designing circuits with one degree of freedom, where the operating point of a block is not firmly set during design but fixed later at the factory production testing stage, usually through component trimming and setting variables in some nonvolatile memory. However, this single degree of freedom, which is limited to a one-time fix, is still not inclusive of all operating conditions. To incorporate multiple fixes, factory calibration is no longer a valid option, and the burden is passed on to the only logical setting: the chip itself. The ultimate solution is assisted-operation whereby a circuit performs its functionality while being mon-itored and continuously steered to its optimal performance by supplying it the appropriate fixes. Assisted-operation is then heavily reliant on on-chip testability to instill awareness of the current circuit performance and operating conditions.

Self-awareness is therefore an inherited product of self-test and a necessary prerequisite for self-calibration. As was mentioned in the earlier chapters, BiST capabilities and techniques are at the core of any assisted circuit designs, especially for RF and analog. In the digital domain, BiST and self-calibration have been successfully applied to provide on-line thermal and power management adjust-ments. The push is to bring that on-line testability and flexibility to the RF and mm-wave domains. Hence, Design-for-Testability and Tunability (DfTT) should become the design method of choice for building firstpass RF SoCs.

RF circuit designs have traditionally opted for analog forms of compensation and feedback to reduce variability, with a multitude of techniques to remove dependency on voltage supply and temperature. However, these methods (mostly biasing circuits) are also implemented as analog circuits and suffer from the same adverse effects as the circuits they are supposedly adjusting. Another way to think of tunability is simply programmability – and what better method to program than in digital. Digital tuning of analog blocks is an interesting alternative that offers a more robust alternative that can replace analog compensation and at the same time ease design requirements.

Analog compensation is a more self-contained method (i.e. a block and its com-pensation circuitry can ideally operate standalone). Digital compensation, on the other hand, requires the presence of a DSP to complete the compensation loop. That requirement is directly met in a radio SoC – and therefore, digital Built-in-Self-Calibration lends itself as a worthy topic of study and development.

5.1.1   Digital Solutions to Analog Impairments

The flexible processing capabilities of the SoC's digital core can be put to use beyond the regular system operation. In the earlier chapters, the digital core is used to control the testing routines and compute different performance metrics. It can be further used to shield the system from failure caused by process and environment conditions. The advantages of the digital approach to monitoring and calibrating RF and analog circuits are many. Adaptive DSP algorithms can be programmed for block- and system-level calibration to achieve performance optimization. DSP solutions take only a fraction of the area of the blocks they are monitoring and would only need to run for a fraction of the time. Calibration routines can run at startup or at predetermined or idle times.

Several existing transceiver blocks can be re-used to make calibration possible, including the ADCs that form the interface to the digital core. The hardware over-head is then limited to the additional sensing and test circuitry. Also, the additional power consumption can be limited only to the time calibration is running and the additional circuits turned off otherwise.

Portability and updatability of the digital approach are also major advantages. Digital circuitry almost always benefits from migration to newer CMOS processes, especially with the relative ease of digital synthesis. Moreover, the calibration algorithms, if implemented in software, can be easily updated allowing an increased level of flexibility, even providing companies with the ability to roll out firmware updates.

5.1.2   Enabling Built-in-Self-Calibration

The mixed-mode nature of RF blocks' digital calibration borrows from the capabilities of built-in-self-test. For calibration to be run, test signals need to be applied to the circuit to be adjusted. The iterative process resembles the setups that can be used for one-tone or two-tone sweeps described in the earlier chapters on BiST. However, instead of dealing with the circuit as a black box and only analyzing its input and output signals, the sweep can also include changes to the circuit conditions. Essentially, this time testing is not used to extract the current performance as its end goal, but only a means to ensuring that the entire set of possible performance parameters is acquired and the best applied.

The self-calibration loop merges BiST and digitally-assisted design principles to force RF blocks into their optimal operating points. It is essentially an iterative process encompassing test, result comparison, and circuit correction. Therefore, looking at the basic loop in Fig. 5.1, the following actions are required at design time:

1. *Enable sensing*: Since the RF blocks' signals are mostly at high frequency, they cannot be readily connected to the digital core for extraction of their properties and hence some type of feature translator is required. The translator takes in an

Fig. 5.1 Self-calibration
loop and its components

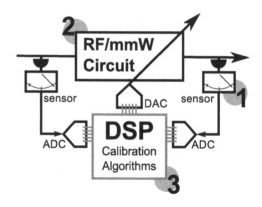

RF signal and outputs a digitally friendly reading. The latter is the easily digitized
and ready for DSP processing. One class of feature translators is discussed in this
book (Chap. 3), the RF amplitude detector. Signal amplitudes can show a direct
or indirect correspondence to circuit parameters and as such can be deconstructed
and interpreted to provide a reading on a desired metric or a comparison between
iterative calibration searches. Other sensing schemes can work at low frequencies to detect nonidealities in the system that are known to definitely present
themselves at RF. An example would be charge pump current mismatch in PLLs.
It may be difficult to detect the resulting increase in phase noise due to that
mismatch (emergence of reference spurs and possible increase in in-band noise).
But it is definitely easier to observe and sense that low frequency mismatch at its
source. Therefore, sensing needs to be enabled in circuits and functional blocks
as part of a self-healing system.

2. *Design digitally-programmable circuits*: The transceiver blocks need to be
   designed with programmable or adjustable elements. The first step is to identify
   the weaknesses of the RF circuit under PVT variations. For example, threshold
   voltage mismatch between differential pairs can be simulated and its contribution to performance degradation quantified. More importantly is finding the most
   effective and suitable correction insertion points. Calibration structures can be
   designed to take on different values corresponding to the correction range and act
   as tuning knobs. The design task is to ensure that the correction range always has
   an optimal point of operation under PVT variations, as seen in Fig. 5.2. There
   may be different access nodes for the correction signals depending on whether
   they affect the operating point of the circuit (usually dc) or are present in the
   high-frequency operation. Circuit biasing is one of the major fixes that can be
   performed as it sets the dc operating point around which most of the ac characteristics center. Therefore, programmable dc biasing in the form of DACs are
   very popular tuning elements, to supply both variable voltages and currents,
   enabling adjustment of the circuit transconductances. Also, ac calibration can be
   implemented in the form of continuous (fine) or discrete (coarse) tuning knobs,
   such as varactors or cap-banks or switchable loads.

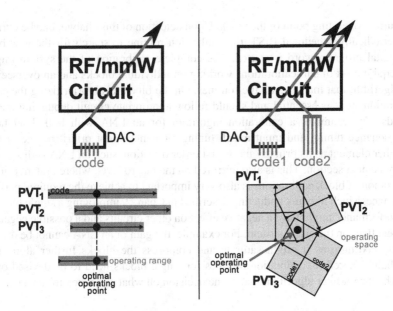

Fig. 5.2 Digitally-assisted RF and mm-wave circuits implemented with wide operation flexibility and ranges always containing the optimal operation point

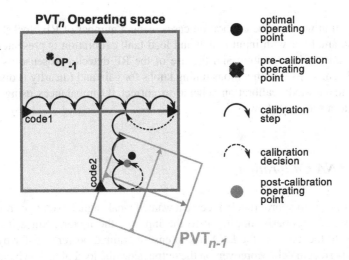

Fig. 5.3 Iterative calibration algorithm with changing PVT conditions

3. *Formulate calibration algorithms*: The cognitive part of the loop resides mainly in the DSP and implemented as calibration algorithms. When the system detects a change in the performance due to PVT variations, calibration can be run to revert the block to an appropriate performance level. This is exemplified in Fig. 5.3 where a change in PVT results in a shift of the operating space and hence the

current operating point of the block. Upon detection of this change, maybe during periodically scheduled BiST, the calibration routine to search for the new best digital program word is started. The complete calibration of the system might require a set of sub-calibrations working on individual blocks and an overseeing algorithm that manages all the routines. On the block level, prioritizing the calibration steps is essential and should follow appropriate circuit debugging methods. For example, a calibration algorithm for an LNA with both load tank resonance tuning and input match tuning elements might prioritize one or the other element as its primary fix or first order of action, such an LNA is discussed in the next section. This is also mirrored on the macro level, where system prioritization of block calibrations is also very important to achieve the required performance. Much like link budgeting described in Chap. 2, improving a single block's performance might have a negative effect on other circuits and a possible negative net effect on the entire system. For example, the gain of an LNA cannot be driven to maximum as it might saturate and compress the blocks further along the chain. Therefore, calibration routines for single blocks need to be devised over which an arbiter algorithm makes the decision on what and when to run each.

5.2   Circuit-Level Tuning

In this section we describe calibration cases using a number of RF blocks. First, a low-noise amplifier with input match and load tank calibration is presented along with its calibration algorithm and the use of the RF detector to sense its performance. Then, a mixer with various tuning knobs for gain and linearity is discussed. Also, a mixed-mode calibration scheme to correct IQ imbalances using the RF detector as a sensor is shown.

5.2.1   LNA Calibration

The LNA needs to appropriately receive a wanted weak signal, isolate it from other interferers, and properly amplify it to the input of the mixers. Since it directly interfaces to the antenna, the LNA has to ensure optimal power transfer by maintaining a perfect match. Moreover, on the output side, the load of narrowband LNAs is a resonant tank offering both gain and filtering centered on the band of interest. However, minor shifts due to parasitics on the input or output of the LNA can shift either or both of them off-center and away from acceptable performance. One of the most popular implementations, the inductively degenerated common-source LNA has an impedance matching network built with purely reactive components but presents a small-signal input impedance comprised of both reactive and resistive elements [1],

$$Z_{in} = j\omega(L_g + L_s) + \frac{1}{j\omega C_{gs}} + g_m \frac{L_s}{C_{gs}} \quad\quad (5.1)$$

where L_g and L_s are the gate and source inductors, C_{gs} is the input device intrinsic gate-source capacitance, and g_m its transconductance. At the resonance frequency, ω_0, given by

$$\omega_0 = \frac{1}{\sqrt{(L_g + L_s)C_{gs}}} \quad\quad (5.2)$$

the input impedance, Z_{in}, is purely resistive and equal to

$$Z_{in} = g_m \frac{L_s}{C_{gs}}. \qu\quad (5.3)$$

To ensure maximal signal transfer, it is important to maintain both the resonant frequency at the desired signal band and the input match quality close to the antenna impedance. Process and temperature variations that alter the passive components' typical values will influence the frequency response of the LNA. The only method to change back these values is to alter the reactive elements. In Ref. [2], the input impedance is changed by trimming the gate inductor, L_g, using switches to short its segments at different locations, effectively creating a variable inductor. This approach is limited in a number of ways, including the need to design and verify custom inductors and the need to mitigate the effects of the finite switch on-resistance on the matching network. Manipulating capacitance is in fact an easier approach to alter reactive values. One widely used example is varactor tuning. In Ref. [3], the authors suggest adding two varactors, one in shunt with C_{gs} and the other with L_s. In their study on a differential 2.4 GHz LNA, they show that the resonance frequency and the input match quality can then be independently adjusted by varying the added varactors' equivalent capacitance through a digitally-programmable bias generator (DAC). Based on the same idea, the load tank can be calibrated to maintain peaking at the required frequency also through varactor tuning of its output inductor [3], and also in Ref. [4].

Furthermore, it is shown that by applying a known signal at the LNA input and observing the output, the amplitude of the latter can provide an indication and measure of the LNA's performance while being calibrated. The best LNA configuration is then the one that achieves the largest amplitude at the output – which is a direct result of a centered peaking frequency response at proper input match and output load resonance. Care should be taken however not to compress the LNA or the circuits following it. Therefore, self-test routines for linearity and compression points, presented earlier in Chap. 4, can set the limits of the calibration.

A 2.4 GHz single-ended LNA based on Ref. [3] is implemented and shown in Fig. 5.4. The LNA is also augmented with the ability to be turned on or off by switching the cascade device's bias between rails. This allows the LNA to be bypassed and shut down as required by other receiver blocks' self-test and self-calibration routines.

Fig. 5.4 LNA with digital
calibration for input match
and output load tuning

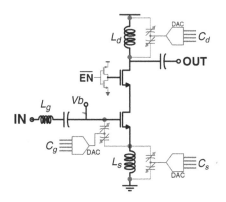

The LNA's own calibration routine then starts by applying a signal from the loopback configuration. The input and output of the LNA are monitored by the RF amplitude detector presented in Chap. 3. The load tank is calibrated first, going through a linear search for the highest amplitude, and hence the lowest dc output of the detector. The optimal code is retained and applied for the next phase of calibration. The same happens at each optimization level. Therefore, the calibration routine passes through the following steps:

1. *Enable loopback*: create one-tone signal in transmitter and loopback to LNA input; monitor both the input signal (to ensure that it is stable) and the output signal of the LNA (the actual observable).
2. *Sweep C_d*: go through the load tank calibration codes ($C_{d,min} - C_{d,max}$) and save the optimal code ($C_{d,opt}$) corresponding to the lowest detector dc value ($dc_{out,min}$). At the end of the search, apply the optimal code and go to the next step.
3. *Sweep C_g*: sweep the calibration codes for shifting the resonance frequency of the input matching network; search for, and apply, $C_{g,opt}$.
4. *Sweep C_s*: sweep the codes for matching quality adjustments; apply $C_{s,opt}$ and end calibration.
5. *Disable loopback*: continue normal operation with new optimal LNA operating point ($C_{d,opt}$, $C_{g,opt}$, $C_{s,opt}$).

An example transient simulation displaying the progression of the calibration algorithm is shown in Fig. 5.5. The RF detector automatically changes modes when it reaches its upper and lower voltage limits; the higher the mode and the lower the dc output of the detector, the larger the amplitude of the signal is. Therefore, the final LNA setting will comprise of the set of varactor words that achieves the lowest possible detector dc output.

The gain of the final calibrated point can be directly extracted from the input and output RF amplitude detectors. These two detectors can keep on monitoring the regular operation of the LNA and if the system detects that the gain has significantly dropped from the last known good value, then calibration can be scheduled to run. Other modifications to the LNA circuit are possible to offer gain adjustment [5] and intermodulation distortion reduction [6].

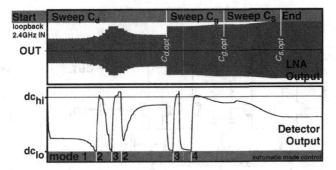

Fig. 5.5 Transient snapshot of LNA calibration routine and the corresponding RF amplitude detector dc output

5.2.2 *Mixer Calibration*

The mixer's gain and linearity are of primary concern in a transceiver chain. The conversion gain in a mixer is defined for signals at different frequencies, at RF and IF, and can be measured using RF and baseband sensors. The intermodulation distortion products that are critical in mixers are not only of the third order (IM3) but also the second (IM2).

Self-test for these metrics is slightly different than that of other RF blocks as the mixer operates at two frequency domains. The RF amplitude detector presented earlier can be used on the RF and LO ports but not on the IF port – at least not for gain and linearity. It can still be placed on the IF port to measure the LO fee through.

An active mixer mainly comprises of three sections: a transconductance part, a switching part, and output loads. The main signal enters the mixer through the transconductance stage, whereas the local oscillator signal acts on the switching stage commutating the signal between the output loads. The various mixer parameters are then determined by one or more of these stages. For example, the mixer's gain is simply a function of its input transconductance and output load. The IIP3 is also a function of the input transconductance. Therefore, adjusting the transconductance (g_m) of the input pair is a key component of manipulating both metrics. On the other end, the main source of IIP2 lies in the switching stage, mostly as threshold mismatch in the different commutating pairs. This mismatch creates an imbalance at the output node. One method to counteract this is to introduce an opposite set of mismatches to counterbalance the output. In Ref. [7], the counter mismatch is introduced as independently switchable output loads at either of the mixer outputs. The net effect is the cancellation of the original threshold mismatch. This technique unfortunately results in dc offsets, a main impairment in direct conversion receivers. The method suggested by Rodriguez et al. [8, 9] implements the counterbalance as independently biased switching pairs, thereby moving one step closer to the actual impediment (V_{th} mismatch) and its effect on the overdrive voltage of the transistors.

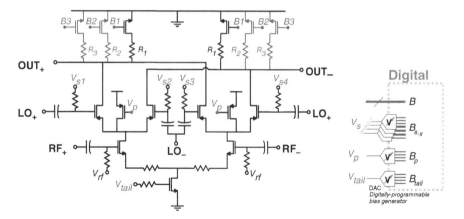

Fig. 5.6 Double-balanced CMOS mixer with digital tuning knobs for gain and linearity calibration [8]

By maintaining an equal overdrive voltage on the switching transistors, the net result resembles threshold-matched identically-biased switches, leading to the severe reduction in IIP2.

The double-balanced CMOS downconversion mixer implementation from Ref. [8] is shown in Fig. 5.6. The circuit's gain adjustment includes transconductance and load tuning knobs. The first is achieved by pumping a variable current into the mixer through either the tail or pMOS current sources, controlled by V_{tail} and V_p, respectively. The second is simply a switched-resistor bank at both output nodes. The IIP2 adjustment knobs are the digitally tunable biases (V_s) at the gates of the switching pairs. A similar idea for IIP2 enhancement for passive mixers is presented in Ref. [10].

For this circuit's calibration, the special case self-test mechanisms for gain and intermodulation distortion of mixers have to be engaged. To calibrate for gain, IIP3 and IIP2, the RF amplitude detector at the RF port is used to obtain the input amplitude whereas a baseband detector measures the resulting IF tones that fall within the low-pass filter bandwidth. When calibrating for gain, an RF one-tone signal is fed back to the mixer from the transmitter via the loopback element. The baseband detector observes the IF signal while the digital gain tuning codes (B_{tail}, B_p, and B in Fig. 5.6) are changed. The search routine for the best set of codes can be done linearly or through a smart calibration engine employing least-mean square (LMS) adaptive algorithms in the DSP block [11, 12]. IIP3 is also calibrated similarly however under a two-tone test while IIP2 is tuned using its own set of digital knobs ($B_{s,x}$ in Fig. 5.6).

5.2.3   IQ Imbalance Calibration

Quadrature error is one of the major impairments affecting the performance of RF systems employing complex modulation schemes. The usual culprit is the local oscillator which fails to supply proper amplitude and quadrature matched signals to drive the mixers in either of the upconversion or downconversion paths.

To achieve good *IQ* balance, designers have opted for layout techniques and new architectures that balance the outputs of the oscillator. However, variations are still largely probable and more so in nanoscale CMOS. An interesting calibration technique relies solely on the digital domain with no direct fix of the actual impairments. The technique, also known as "dirty-RF" [13], posits that knowledge of a transceiver's analog impairments can be countered purely in the digital domain. That is, knowing the *IQ* imbalance of the transmit chain, a pre-distorted symbol is transmitted from baseband such that the introduced distortion and the inherent RF imbalance cancel out and a clean symbol is transmitted. The same also applies for receivers, where post-distortion of the received skewed symbols renders them ideal.

The first step is then to detect these imbalances. We have shown in Chap. 4 how using the detector to sense the differential quadrature outputs of the LO can enable amplitude mismatch detection and phase imbalance prediction accurate to less than a degree (see Fig. 4.9). A hybrid mixed-mode detection and calibration technique can then be implemented, making use of the accuracy of the amplitude and phase mismatch detection with digitally-assisted analog compensation and DSP signal correction.

Using the amplitude and phase mismatch information obtained from self-test, the calibration algorithm in the DSP can correct for these mismatches in a number of possible ways. Several digitally-controlled analog correction knobs can be used to correct for the amplitude mismatch: LO phase-shift element tuning, mixer gain tuning (e.g. the mixer presented in the previous section), or alternatively tuning in baseband at the VGAs after (or before) the mixer. Once the gains of the *I* and *Q* paths are equalized, the phase mismatch remains.

Considering a demodulator input cosine signal at RF (f_{RF}) and phase mismatched (φ) LO signals, the sampled *I* and *Q* signals after the low-pass filter are

$$I[n] = \frac{1}{2}\cos(2\pi f_{IF}n) + d_i$$

$$Q[n] = \frac{1}{2}\sin(2\pi f_{IF}n - \varphi) + d_q = \frac{1}{2}\sin\left(2\pi f_{IF}(n-r)\right) + d_q \qquad (5.4)$$

where $r = (\varphi/2\pi f_{IF})$ is a fractional time delay, and d_i and d_q are dc mismatch terms [14]. The latter dc terms can be readily removed by the DSP. However, to complete the compensation, a fractional delay is needed on one of the channels. Fractional delay digital filters can hence be used to implement the phase calibration engine. One popular implementation of fractional-delay filters is the Farrow filter [15]. The Farrow filter is a time-varying FIR filter with fixed filter coefficients and a programmable delay component. Its structure is shown in Fig. 5.7 with *r* as the fractional delay. It is essentially an interpolator that generates samples at fractions of the sampling period and used in a number of diverse signal processing applications. It possesses many advantages including its guaranteed stability (as it is a feed forward architecture) and also low hardware complexity. Techniques to further reduce its hardware complexity have been devised to implement its coefficients as sum-of-products-of-two (SOPOT) therefore greatly easing the multiplication burden [16, 17].

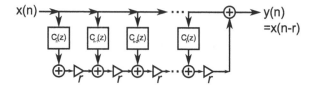

Fig. 5.7 Basic farrow filter structure

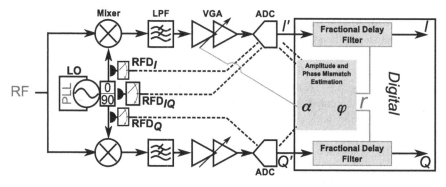

Fig. 5.8 Mixed-mode IQ imbalance compensation

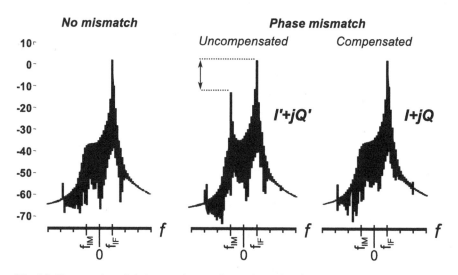

Fig. 5.9 Test tone demodulation: quadrature phase mismatch before and after compensation

Putting it all together, Fig. 5.8 shows the mixed-mode calibration setup for the receiver. A similar setup is possible for the transmitter pre-distortion. The IQ imbalances are extracted from the RF amplitude detector signatures at the LO. The amplitude mismatch is corrected by adapting the VGA of one of the paths whereas the phase mismatch is compensated in the digital fractional-delay filters. Figure 5.9 shows an example of the effect of phase mismatch on the emergence of an image tone and its subsequent suppression by the Farrow filter in the digital baseband.

5.3 Summary

This chapter explored the built-in-self-calibration of RF transceivers and more importantly its requirements. In an RF SoC, the digital domain can be leveraged to compensate for analog and RF nonidealities by either correcting them on the circuit level or on the algorithmic level. The strength of the digital parts of the chip and the capabilities that can be embedded in them allows for robustness enhancement measures to be put in place. This robustness is a product of self-aware blocks and systems that are capable of monitoring their own performance and mending its weaknesses. Self-test is therefore a vital part of calibration as it provides the metrics on which a calibration routine operates. The digital self-calibration loop is therefore enabled with efficient sensing, tunable circuits, and DSP algorithms. Examples of tunable circuits are shown with their calibration schemes using the self-test methodologies presented in earlier chapters.

References

1. B. Razavi, *RF Microelectronics*, 1st ed. Upper Saddle River, NJ: Prentice Hall, 1997
2. T. Das, A. Gopalan C. Washburn, P.R. Mukund, "Self-calibration of input-match in RF front-end circuitry," *IEEE Transactions on Circuits and Systems II*, vol.52, pp. 821–825, Dec. 2005
3. J. Wilson, M. Ismail, "Input match and load tank digital calibration of an inductively degenerated CMOS LNA," *Integration, the VLSI Journal*, vol. 42, no. 1, pp. 3–9, Jan. 2009
4. N. Ahsan, J. Dabrowski, A. Ouacha, "A self-tuning technique for optimization of dual band LNA," *EuWiT 2008. European Conference on Wireless Technology, 2008*, pp.178–181, Oct. 2008
5. C.-H. Liao, H.-R. Chuang, "A 5.7-GHz 0.18-μm CMOS gain-controlled differential LNA with current reuse for WLAN receiver," *Microwave and Wireless Components Letters, IEEE*, vol.13, no.12, pp. 526–528, Dec. 2003
6. T.-S. Kim, B.-S. Kim, "Post-linearization of cascode CMOS low noise amplifier using folded PMOS IMD sinker," *Microwave and Wireless Components Letters, IEEE*, vol.16, no.4, pp. 182–184, April 2006
7. K. Kivekas, A. Parssinen, J. Ryynanen, J. Jussila, "Calibration techniques of active BiCMOS mixers," *IEEE Journal of Solid-State Circuits*, vol.37, no.6, pp.766–769, Jun 2002
8. S. Rodriguez, A. Rusu, L.-R. Zheng, M. Ismail, "Digital calibration of gain and linearity in a CMOS RF mixer," *ISCAS 2008. IEEE International Symposium on Circuits and Systems, 2008*, pp.1288–1291, May 2008
9. S. Rodriguez, A. Rusu, L.-R. Zheng, M. Ismail, "CMOS RF mixer with digitally enhanced IIP2," *Electronics Letters*, vol.44, no.2, pp.121–122, Jan. 2008
10. S. Rodriguez, S. Tao, M. Ismail, A. Rusu, "An IIP2 digital calibration technique for passive CMOS down-converters," *Proceedings of 2010 IEEE International Symposium on Circuits and Systems (ISCAS)*, pp.825–828, June 2010

11. S. Rodriguez, A. Rusu, L.-R. Zheng, M. Ismail, "A Novel BiST and Calibration Technique for CMOS Down-Converters," *ICCSC 2008. 4th IEEE International Conference on Circuits and Systems for Communications, 2008*, pp.828–832, May 2008

12. K. Dufrene, Z. Boos, R. Weigel, "Digital Adaptive IIP2 Calibration Scheme for CMOS Downconversion Mixers," *IEEE Journal of Solid-State Circuits*, vol.43, no.11, pp.2434–2445, Nov. 2008

13. G. Fettweis, M. Löhning, D. Petrovic, M. Windisch, P. Zillmann, W. Rave, "Dirty RF: A New Paradigm," *International Journal of Wireless Information Networks*, vol.14, no.2, pp.133–148, June 2007

14. H. Wang, Y. Lu, X. Wang, C. Wang, "Digital I/Q Imbalance Compensation in Quadrature Receivers," *CIE '06. International Conference on Radar, 2006*, pp.1–4, Oct. 2006

15. C.W. Farrow, "A continuously variable digital delay element," *IEEE International Symposium on Circuits and Systems, 1988*, pp.2641–2645 vol.3, Jun 1988

16. C.K.S. Pun, Y.C. Wu, S.C. Chan, K.L. Ho, "On the design and efficient implementation of the Farrow structure," *Signal Processing Letters, IEEE*, vol.10, no.7, pp. 189–192, July 2003

17. J. Yli-Kaakinen, T. Saramaki, "Multiplication-Free Polynomial-Based FIR Filters with an Adjustable Fractional Delay," *Circuits, Systems, and Signal Processing*, vol.25, no.2, pp.265–294, Apr 2006

Chapter 6
Conclusions

The technology breakthroughs resulting in the successive scaling of CMOS transistors has kept Moore's law alive for more than half a century. An important effect of physical dimension reduction is the increase in attainable operating frequencies. Therefore, devices are not only smaller but also faster, fast enough to surpass the RF spectrum and even the millimeter-wave region. This allowed CMOS to contend for integrated circuits for wireless connectivity applications. With the proliferation and surge of demand for ubiquitous computing and connectivity, the customer pool has expanded tremendously making wireless (devices, applications, services, etc.) one of the fastest growing markets worldwide. The new challenge then becomes to take the cost-effective and mass-production-ready CMOS beyond memory and logic into the "More than Moore" regime. This represents a new dimension in integration, whereby hybrid mixed-mode systems co-exist on the same silicon substrate, essentially combining memory, logic, power, analog, and RF circuits into one: a radio System-on-Chip.

Integrating RF and mm-wave capabilities with powerful digital processing offers a great deal of possibilities, if only it were an easy task! The challenges turn out to be as great as these possibilities. The inherent increase in variability of nanometer CMOS device performance translates into extreme variations of circuit functional metrics. Coupled with the tighter requirements for new applications and standards, the percentage of working parts that also pass the specifications greatly diminishes. The yield problem is more pronounced in RF SoCs as the radio frequency circuits that occupy only a part of the area are much more vulnerable to failure thereby setting the passable limits of the entire system. The seemingly unwieldy RF circuits do have an optimal operating region; the only downside is the narrowness of that region making minor shifts in operating conditions manifest as extreme roll-offs in performance. However, that operating region can be still recuperated – although demanding direct intervention. To regain lost performance, the first step is to know what the current state of operation is. Therefore, testing is the first step towards obtaining this knowledge. Techniques for testing RF circuits are then of primary importance. External testing is quite prohibitive, time- and cost-wise, especially in

S. Bou-Sleiman and M. Ismail, *Built-in-Self-Test and Digital Self-Calibration for RF SoCs*, SpringerBriefs in Electrical and Computer Engineering, DOI 10.1007/978-1-4419-9548-3_6, © Springer Science+Business Media, LLC 2012

highly complex systems employing a mix of digital, analog, mixed-signal, and RF. Design-for-Test provides a partial solution to this difficulty by enabling test structure on-chip embedding, for increased visibility and observability. However, external testers are still required which limits testing and verification to the lab.

In this book, we highlighted the requirements for migrating the test completely on-chip and allowing it to be portable. The portability property carries with it some interesting possibilities of testing a system on-the-fly and while in actual operation, where changes in operating conditions are the most probable and representative. This equates to building functional blocks in conjunction with testing blocks at the very onset of the design process. However, the desire is to include more functionality and not more testability, putting a reasonable limit on the overhead associated with on-chip test. To this end, we propose a test sensor that lends itself to accurate parametric extraction of various RF and mm-wave blocks and, at the same time, to non-invasive and minimal-overhead integration. The sensor, an RF amplitude detector, translates a high frequency signal into a corresponding low frequency or dc reading. That reading can be easily assessed by the digital parts of the SoC and helps in quantifying a signal's properties and a block's performance. The proposed RF detector benefits from high conversion gain (sensitivity) and wide dynamic range (extent of amplitude detection) while being frequency non-specific. Three implementations are compared to available solutions in literature whereby our designs position themselves as front contenders for true on-chip BiST. An enhanced loopback BiST architecture is also proposed making use of a signal loopback element and multiple sensing nodes to perform a number of parametric tests. The complete system should be able to locally generate test signals, sense the response to these signals, and interpret the signatures of the sensing results. These steps are all possible within the boundaries of an SoC, and with very little overhead. The RF amplitude detector links the RF and digital at its ends.

Several block metrics can then be inferred directly or indirectly from simple amplitude measurements. The test setups are discussed for a number of blocks from LNAs, mixers, local oscillator, and quadrature modulators and demodulators. Methods to quantify gain, linearity, intermodulation distortion, isolation, and quadrature mismatch are explained and example cases demonstrated for both micro- and millimeter-wave spectrum.

The ultimate goal is not only test but actual performance enhancement. There is a certain limit to making an RF circuit rigid and highly stable. This limit has been pushed to extents demanding unreasonable design effort while still not guaranteeing proper functionality across all cases. Surprisingly, the answer is exactly the opposite: flexibility. Flexible RF circuits are not overdesigned to perform at their best operating point but actually loosely designed with enough wiggle room. The margin is relaxed in anticipation of assisted-operation, i.e. an active mechanism will regulate the operating point to match the operating conditions resulting in performance enhancement. The testing techniques we presented will enable the system to become self-aware and therefore know its location within the operating margin and work its way towards the optimal point inside that margin. With flexible digitally-assisted RF circuits, the optimization can run on the more robust and highly-efficient digital parts.

A few examples for digital tunability of RF circuits are presented including LNA, mixer, and quadrature modulator. The highlighted techniques and test strategies can then be extended to a number of other circuits.

With the newer CMOS processes showing no sign of variability restraint, a successful self-calibration architecture is tantamount to a successful chip implementation. Therefore, the importance of self-test and self-calibration for RF SoCs cannot be underestimated and has to be employed as a new multifaceted design paradigm employing techniques and best practices from across the board.